AGAINST THE TIDE

AGAINST THE TIDE

The Battle for America's Beaches

Cornelia Dean

COLUMBIA UNIVERSITY PRESS

NEW YORK

Columbia University Press
Publishers Since 1893
New York   Chichester, West Sussex
Copyright © 1999 Cornelia Dean
All rights reserved
Library of Congress Cataloging-in-Publication Data
Dean, Cornelia.
Against the tide : the battle for America's beaches / Cornelia Dean.
p.   cm.
Includes bibliographical references and index.
ISBN 0–231–08418–8
1. Coast changes—United States. 2. Beach erosion—United States.
3. Coastal zone management—United States. I. Title.
GB460.A2 D4   1999
333.91'715'0973—dc21   98–50755

Casebound editions of Columbia University Press books are
printed on permanent and durable acid-free paper.
Printed in the United States of America
c 10 9 8 7 6 5 4 3 2

In memory of my father,
Joseph L. Dean

CONTENTS

Almost twenty years ago, I fell in love with a beach, a barrier island at the mouth of a small river, on the coast of southern New England. The island lay between the crowded villages of Cape Cod and the glitz of Newport, but somehow development had passed it by.

A friend introduced me to this beach one summer day. Ignoring the state beach concession area and a town parking lot, we drove into a boatyard that occupied a patch of dry ground on the island's marshy back side. We made our way through the yard, past wooden skiffs and the occasional sloop, hauled out and lying in wooden cradles. In the water, catboats nodded at their moorings; grimy working boats were tied up at a fishing pier. We parked at the back of the yard, behind the gray plank dry dock sheds, and set out along a narrow sandy path. To the left, dunes rose behind the stunted pines and scrub oak. On the right, undulating marsh grass spread like a carpet toward the estuary glimmering in the sun. Poison ivy snatched at our bare ankles, and mosquitoes and greenflies dived at us as we struggled in the deepening sand. But when the trail finally rose to cross the dunes, we had our reward: a secluded stretch of beach where the sand was clean and soft, the air was filled with the tang of seaweed, and the water was so clear it was almost possible to count the grains of sand moving in languid currents along the bottom.

Over the years, I would learn the ways of this beach—the coming and going of the wave-tossed red seaweed, the changing currents that carried slipper shells on shore one week, quahog shells the next. One day, I saw hundreds of tiny hermit crabs, no longer than an inch or two, march in ranks around the end of the island into the marsh behind it, driven by some mysterious biological imperative.

Like all beaches, this beach was never exactly the same from one day to the next. The currents, the tides, the storms were constantly obliterating its features, only to re-create them in slightly different shapes. The sand that collected around my feet as I waded in the surf one day would be gone the next, carried by winds into the dunes, or by storm waves into an offshore bar. As one grain was carried away, another came along to take its place.

Years later, after I had left New England and lived in New York City for almost a decade, I revisited the beach. Someone, perhaps the proprietor of the boatyard, had moved a few granite boulders to block the path my friend and I had taken. But it was easy to climb around them and, though the trail had narrowed somewhat, the walk past the foliage and marsh was as beautiful as ever. The beach itself was pristine, as it had ever been.

In researching this book, though, I learned that my beach had not always been so innocent of development. Around the turn of the twentieth century it was a summer resort, a vibrant colony of elaborate shingled "cottages" on the dunes. Wealthy people came from as far away as Philadelphia to escape city heat and enjoy ocean breezes.

All that changed on the first Sunday in September 1938, when, with almost no warning, a hurricane roared out of the Atlantic, cut a devastating swath across Long Island, and blasted its way into New England, pushing a twenty-foot storm surge before it. Wind and water swept everything from the beach, even the dunes. One of the few houses that survived the initial blast was last seen heading inland, riding the pounding surf.

When the survivors returned to the beach the next day, they did not recognize the landscape. "My breath was taken away and my heart virtually stopped," one of them recalled decades later. "What I saw was a wasteland of sand and stone, with not one stick left of the scores and scores of buildings that lined the shore. Even the paved road was gone. I could see that the hurricane-whipped seas had swept away everything."[1]

It took years for nature to rebuild the beach, but it did. Meanwhile, state and local agencies bought the land, preventing people from building on it again. Today, the rippling dunes and expanses of sand are impervious to storms and erosion. That is not to say storms and erosion don't eat away at them. Far from it. Erosion rates in this area are as much as three feet a year, and the coastal storms in New England winters are as fierce as ever. But because the beach has few buildings on it, it is free to shift as water rises and storms come and go. Like most natural beaches, it can defend itself against bad weather. Its position may change, but it survives.

Such is not the case with most of the rest of the American coastline. Atlantic, Pacific, and Gulf coast beaches are sinking under rising sea levels and floods of development. Pinned down by people and their buildings, beaches are drowning in place. Plenty of remedies are offered to stem this beach loss—everything from rock walls to discarded Christmas trees. Most do more harm than good.

In the long run, extensive development cannot coexist with an eroding beach—and most American beaches are eroding. We Americans must reconsider our attitude toward our beaches. That is the reason for this book.

Chappaquiddick, 1999

This is a journalist's book, not a scientist's or a scholar's. It is an impressionistic tale woven of string gathered in interviews, at scientific meetings, on field trips, and in conversations over a decade of travel along the East, West, and Gulf coasts. It is not definitive. Too much is happening—too quickly and in too many places—for it to be so. But I hope that it will help readers better understand the struggle under way now for the future of the American coast.

Many people helped me with this book. First among them are the scores of coastal scientists and engineers who have answered my questions, allowed me to accompany them in the field, given me copies of their papers, and pointed me toward other sources of information. This book would have been impossible without their help.

Among them are: David G. Aubrey, Willard Bascom, Kevin R. Bodge, Charles A. Bookman, Dave Bush, Nicholas Coch, Richard Creter, Robert G. Dean, Kathie Dixon, Robert Dolan, Bruce Douglas, Reinhard E. Flick, Arthur Gaines, Graham Geise, James W. Good, Gary Griggs, Mark Hay, Rob Holman, Peter Howd, Douglas Inman, Chris Jones, Timothy W. Kana, Joseph Kelley, Paul Komar, Steve Leatherman, Virginia Lee, Robert H. Osborne, Randall W. Parkinson, Shea Penland, Bernard W. Pipkin, Orrin H. Pilkey, Jr., Norbert Psuty, Stan Riggs, Harry H. Roberts, Dan Rubenstein, S. P. Schiff, Bob Sheets, David Skelly, David Soileau, Tom Terich, Bob Wiegel, Jeff Williams, and Carl Zimmerman.

I also owe many thanks to the staff of the Army Corps of Engineers, especially the scientists, engineers, and others at the Waterways Experiment Station in Vicksburg, Mississippi, and, in particular, to William R. Birkemeier and his colleagues at the Corps' Field Research Facility in Duck, North Carolina. They have been more than generous with their time and expertise.

I must also thank the DeWitt Wallace Foundation and its fellowship program for journalists, which enabled me to spend a month at the Duke

University Marine Lab in Beaufort, North Carolina, whose scientists and staff members, particularly Joe Ramus and Cindy Baldwin Adams, were extremely helpful.

A number of state and municipal officials, land use experts, civic leaders, developers, and coastal residents also provided much valuable information and assistance. Among them are David Belcher, Daniel V. Besse, Bruce Bortz, Donald W. Bryan, David Brower, Mike Buckley, Armand Cantini, Barbara Krantz Crews, Mark Crowell, Sally S. Davenport, Todd Davison, Wynne Dough, Russell Eitel, James E. Fauzs, Peter H. F. Graber, Jack Gray, Wayne Gray, Camilla Herlevich, Christopher Kennedy, Michael Kennedy, David Lucas, Orville and Karen Magoon, James M. McCloy, Beth Millemann, Todd Miller, Bob Morton, David Owens, George Owens, Lena Ritter, Harry Schiffman, Bob and Marlene Seaborn, Rich Shaw, Jeff and Darragh Simon, Ken Smith, Gerry Stoddard, Katherine Stone, Stan Tait, Virginia K. Tippie, Gary A. Vegliante, Gregory Woodell, and officials of the Port Canaveral commission.

I am grateful also to the librarians and research assistants who helped at the Rosenberg Collection of the Galveston Library, the public libraries of Westport and Chatham, Massachusetts, the Woods Hole Oceanographic Institution, the Supreme Court of the United States, and the *Vineyard Gazette* in Edgartown, Massachusetts, where Eulalie Regan was especially helpful.

Finally, I must thank friends and family members who encouraged me in this long-running project, especially my mother, Winifred Dean, and my sister, Barbara Dean, and those who offered accommodation and other assistance: Sandra Blakeslee, Jonah Freedman, Karl Manheim, Catherine W. O'Malley, T. J. and Anne O'Malley, Sallie Riggs and Mike Corrigan, and Harold and Lois Schmeck. Ed Lugenbeel of Columbia University Press had faith in this project when many others did not, and his colleague Holly Hodder brought it to fruition, along with Jonathan Slutsky, Anne McCoy, and Susan Heath.

Many others assisted me along the way. I have tried to do justice to the help and information all of these people have provided me. If there are errors in this book, they are mine.

AGAINST THE TIDE

CHAPTER ONE

September, Remember

> June, too soon.
> July, stand by.
> August, look out you must.
> September, remember.
> October, all over.
> —R. Inwards, *Weather Lore*

As the twentieth century dawned on Galveston, Texas, the island city was a thriving place. Built on a narrow, sandy barrier island at the mouth of Galveston Bay, the city had a fine anchorage and a three-mile railroad causeway to the mainland, and together they made Galveston the center of commerce for the entire Southwest. Nearly all the goods and people entering Texas came across the Gulf of Mexico to Galveston, and it was fast becoming the number one cotton port in the world. Its forty thousand residents enjoyed two opera houses and the benefits of a recently established medical school. Galveston's wealth, per capita, was among the highest of any city's in the country, and its merchants, bankers, and builders displayed it in elaborate brick and timber mansions along the city's stately avenues. Along its side streets, frame houses or small apartment blocks housed the city's growing middle class. On fine days in the summer of 1900, Galvestonians strolled along an oceanfront promenade or took their ease on the verandas of the resort hotels set back from their wide sand beach.

As on most days that summer, the *Galveston Daily News* of Saturday, September 8, held little that was startling: diplomats in China were negotiating over the Boxer Rebellion; a survey of the Brazos River was progressing nicely; the price of cotton was steady on the Liverpool market. On page 2, a small item noted: "Storm on Florida Coast." A storm had struck the east coast of Florida, the article reported, and had moved across the state and into the Gulf of Mexico. "The gale continues," it said.

The next day's edition told a different story. Printed with obvious difficulty on narrow paper, it said, in its entirety:

GALVESTON NEWS
Sunday, Sept. 9, 1900
Following is list of dead as accurately as News men have been able to make it. Those who have lost relatives should report same at News office. This list will be corrected and added to as returns come in.

Below were two columns of names—men and women, husbands and wives, entire families. Everyone reading the paper knew what had killed them. The storm barely worth noting the day before had strengthened into a hurricane, crossed the Gulf, and struck a mortal blow at Galveston. Overnight, the city had lost 20 percent of its people in what remains the deadliest natural disaster ever in the United States.

Isaac Monroe Cline was in charge of the U.S. Weather Service office in Galveston that Saturday. Though he had worked for the Weather Service for eighteen years, eleven of them in Galveston, he had never seen a hurricane, and he had received only fragmentary ship reports of the storm. Still, as the wind stiffened that morning, the tide rose far above normal, and ocean swells grew in height and frequency, Cline knew a severe storm was bearing down on the island. Without waiting for the obligatory instructions from Washington, he ordered the hurricane flags hoisted, and he set out to warn as many of his fellow citizens as he could.

"I harnessed my horse to a two wheeled cart which I used for hunting and drove along the beach from one end of the town to the other," he wrote decades later. "I warned the people that great danger threatened them."[1]

Cline said later that his warning had encouraged holiday-makers to cut their visits short and board ferries or trains for the mainland. But even on a calm day, the railroad trestle barely cleared the waters of Galveston Bay, and this day was no longer calm. By noon, well before the storm hit, the bridge was swamped and ferry service to the mainland had halted. Galveston was cut off.

Ideally situated for business, the city was a bad place to be in a hurricane. Galveston Island is little more than a thirty-three-mile ribbon of sand, only two miles wide at its widest. The city's highest point was not even nine feet above sea level,[2] its average elevation only about five feet. Even that had been achieved at a price: sand dunes that might have offered some protection from the storm had been taken down and used to fill low places in the city's business district.

By midafternoon, the entire island was under water. Cline dispatched his brother and assistant, Joseph Cline, to telegraph the chief of the Weather Bureau in Washington that "an awful disaster was upon us."[3] Wading through water up to his waist, the younger Cline made it to the telegraph office, only to find the lines were out. One telephone line was working intermittently; just as he got his message to the Western Union office in Houston, it failed for good.

Joseph Cline returned to the Weather Service office, and he and his brother decided to make for home. Struggling against wind, rain, and still-rising water, they waded almost two miles to Isaac Cline's house, where more than fifty people, including his wife and children, had sheltered. With storms in mind, he had built a stout house, and it stood for hours as the wind grew stronger and waves sent the debris of more fragile structures crashing against it. No one knows for sure how fast the wind blew—later, meteorologists estimated that sustained winds reached 120 miles per hour at the height of the storm.

As the water rose, as high as sixteen feet in the heart of the city, the refugees in the Cline house moved to the second floor and, for a while, it seemed the house and its occupants might survive. But about 8 P.M., as lightning shattered the darkness, Isaac Cline looked out a window and saw part of a streetcar trestle, torn loose by the waves, heading toward him like a battering ram. The house was wrecked and everyone inside was thrown into the storm.

Isaac Cline, his three children, Joseph Cline, and another little girl they snatched out of the water spent the next four hours clinging to floating wreckage and listening to the wind and the shrieks of the injured. By the time the waters began to recede around midnight, nearly everyone else who had sheltered in the Cline house was dead. "My wife's clothing was entangled in the wreckage and she never rose from the water," Cline recalled. Her body was discovered several weeks later, tangled in the debris her family had clung to through the night.[4]

When the sun rose Sunday morning, six thousand people or more—no one could say exactly how many—had been drowned or battered to death in Galveston. House after house, block after block had been reduced to rubble. Streets were impassable, and even substantial public buildings were badly damaged.

"The city of Galveston is wrapped in sackcloth and ashes," the Associated Press reported. "She sits beside her unnumbered dead and refuses to be comforted. Her sorrow and suffering are beyond description. Her grief is unspeakable."[5]

The dead lay where the retreating storm had left them. There was no way and no place to bury the bodies, so they were loaded on barges and dumped at sea, only to float to shore with each new tide. In desperation, the military authorities who took over the city ordered that the corpses be burned, in beach bonfires fueled with tons of wreckage. This work, so gruesome that men had to be forced at gunpoint to perform it, went on for weeks; the fires could be seen from the mainland until November. Eventually, the *Galveston Daily News* listed 4,263 identified dead; it is estimated that an additional 1,500 bodies had been disposed of without identification.

Wreckage, bodies of the dead tangled within it, filled Galveston streets when the storm had passed. To raise relief funds, Clara Barton, head of the American Red Cross, organized the sale of photos of the disaster, including this one. *(Rosenberg Collection)*

GALVESTON NEWS.

SUNDAY, SEPT. 9, 1900.

Following is list of dead as accurately as we have been able to make it. Those who have lost relatives should report same at News office. This list will be corrected and added to as returns come in.

Amundsen, Mrs.
Mrs. Aigily.
Allen, Mr. and Mrs.
Alberts, M. wife and daughter.
Betts, Walter.
Broecker, Mr. and Mrs. John F. and two children.
Bergstedge, Mrs. and two children.
Bird, family of officer.
Barlows, Mrs.
Baxter, Mrs. and child.
Bell, Mrs. Dudley
Bird, Mrs. and child.
Barnett, Mrs. George and child.
Mrs. Broecker.
Bowe, Mrs. John and two children.
Benn, Mrs. Annie and two naughters.
Barnett, Mrs. Gary and two children.
Bowe, Mrs. John and three children.
Burns, Mrs.
Compton, Mr. and Mrs. A. J.
Cramer, Miss Bessie.
Coryell, Patti Rose.
Cline, wife of Dr. I. M.
Clark, Mrs. C. T. and child.
Charles Davies.
Coryell, Mrs. J. R.
Collins, daughter of Mr. Irs.
Coates, Mrs. William A.
Clark, Mrs. C. T. and child.
Caddou, Alex. and five children.
Mrs. John and two children, Compton.
Dailey, William
Day, Alf.
Delaney, Mrs. Jack and two children
Delany, Capt. Jack, wife and son.
Davis, Gussie.
Dorin, Mrs.
Dorin, Miss Jennie.
Delay, Paul.
Davenport 3 children of Mr. and Mrs.
Dorian, Mrs. and five children.
Davies, John R. and wife.
Ewing, Miss
Engelke, John, wife and child
Exnes, Mrs. Katy and two daughters.
Ellmer, two children of Capt. Will.
Eichler, Edward.
Fredrickson, Viola.
Fredrickson, Mrs. and baby.
Fialer, Mr. and Mrs. Walter E. and two children.
Fordtram, Mrs. Claud G.
Flash, Mr. William.
Foster, Mr. and Mrs. Harry and three children.
Guest, Mamie.
Gordon, Miss.
Gordon, Mrs. Abe and three children.
Gordon, Miss.
Garmud, John H. wife and two children.
Gernard, Mrs. John and two daughters.
Hughes, Mrs.
Howth, Mrs. Clarence.
Hill, Mrs. Ben and child.
Harris, Mrs. J. H.
Harris, Miss Rebecca.
Harris, Mrs. (colored)
Holbeck—and boy.
Humburg, Mrs. Peter and four children.
Hawkins, Mattie Lee.
Huhn, Mr. F.
Hausinger, Mr. H. A. daughter and mother-in-law.
Hughs, Joe.
Herr, Miss Irene.
Howe, Adolph, wife and five children.
Jones, Mr. and Mrs. and daughter.
Johnson, Richard (Colored)
Jones, Mrs. W. R. and child.
Kelner, Sr. Charles L.
Kelly, Barney.
Kelly, Willie.
Krauss, Miss Kate.
Lauterdale, Mrs. Robert, two daughters, one son and Mrs. Lauderdale's mother.
Lisbony, W. H.
Lablatt, Joe.
Lynch, John.
Lord, Richard.
Lafayette, Mrs. and two children.
Longnecker, Mrs. A.
Lenker, Tommy.
Lasseco, Mrs.

Love, R. A. officer.
Magus, Mr. two daughters and son
McCauley, Annie.
McKenna, five members families of P. J. and J. P.
Monroe (colored), Mrs. and three children.
Masterson, B. T. and family.
Munn, Sr., Mrs. J. W.
Munn, Mrs. J. W.
Motter, Mrs. and two daughters.
Norton, Mrs. and two children.
Nolley, Mrs. Sam and four children.
O'Keefe, Mrs. Mike and brother.
O'Herrow, Wm.
O'Dell, Miss Nellie and brother.
Palmer, Mrs. J. B. and child.
Parker, Miss Mollie.
Plitt, Harmon.
Peek, Capt. R. H., wife and 5 children.
Posette, Josephine.
Parker, Mrs. Frank and two children.
Ptolmey, Paul.
Parker, Angeline.
Parker, Sullivan, wife and 3 children.
Porter, Henry.
Pitz, C. H.
Quester, Bessie.
Quester, Mrs. M. son and daughter.
Rust, Charles.
Roudadaux, Murray.
Rice, Will J.
Roll, J. F., wife and four children.
Ripley, Henry.
Kegan, Mike, wife and mother-in-law.
Ross, Mrs. Franklin.
Rhymes, Mr. Thomas, wife and 2 children.
Rohter, Albert and wife.
Richards, officer.
Rowan, family of officer.
Swain, Richard D.
Spencer, Stanley G.
Stieglock, Miss Mabel
Spanish sailor; Steamship Telesfora
Scofield, Miss Ida
Sommers, Miss Helen.
Swelgal, George, mother and sister.
Schwartzbach, Joe.
Sherwood, Charles.
Smith, Mrs. Mamie.
Schuler, Mr. and Mrs. Charles and five children.
Sharp, Mr. and Mrs.
Sharp, Miss Annie.
Schuize, Mr. and Mrs.
Summers, Sarah.
Sylvester, Miss.
Schroeder, Mrs. George M. and four children.
Schuler, Adolph, mother and five sisters.
Treadwell, Mrs. J. B. and child.
Taylor, (colored), Mrs.
Trebosius, Mrs. George.
Trebosius, two sisters of George Trebosius.
Watkins, Mr. S.
Wenamore, family of seven members
Wenman, Mrs. John C. and two children
Wilson, Mrs. B.
Webster, Edward and two sisters.
Woodward, Miss Hattie.
Woollam, C.
Wilson, Mary and child.
Wallace—, and four children
Wood, mother of deputy U. S. Marshal
Webster, Sr. Thomas.
Wolfe, Charles.
Walter, Mrs. Charles and three children.
Williams, Miss.
Weinberger, Mrs. Frederick.
Weinberger, Fritz J.
Wharton.
Warren, James, wife and 6 children.
Wood, Mrs. S. W.
Woodward, Mrs. R. L. and two children.
Wakster, Mrs. David.

UNACCOUNTED FOR AND MISSING:

Bergman, Mrs. R. J. and little daughter, County, Tenn.
Edwards, A. R. G. and family,
Lablatt, and daughter,
Masterson,
Minns, Sr.
Plitt, officer Police
Rhea Mrs. and Miss Mamie Rhea of Giles
Withey, H.

The *Galveston News*, all of it, the day after. *(Rosenberg Collection)*

As soon as they were able, some survivors packed up what they could salvage and abandoned Galveston for the mainland; others thought the entire city should move inland. But Galveston's leaders were determined to stay—and determined that such horror would never touch them again.

They devised a plan that even in an era of engineering daring stood out for its size, cost, and audacity. They would wall off their city with a three-mile concrete barrier, seventeen feet high, sixteen feet wide at the base, and five feet wide at its top. Then, they would raise the elevation of the city behind it by as much as seventeen feet, out of reach of any storm. With confidence—or bravado—Galvestonians declared it would be the greatest engineering feat since the construction of the pyramids.

Money was the project's first big obstacle. For one thing, though the shipping business quickly began to recover, much of the city's tax base had been destroyed, along with many of its roads, utility lines, and other infrastructure. Nevertheless, plans were drawn up and advertised, and a seawall bond referendum was held in November 1902. The measure passed by a vote of 3,085 to 21. Construction began immediately, and twenty-one months later the wall was finished.

Meanwhile, engineers determined what the grade would have to be to meet the wall, and on every street workers painted white lines on utility poles to mark the height. Owners of buildings that had survived the storm were given a choice: jack them up, tear them down, or see them drowned in silt.

To deliver the fill, engineers dug a canal down the middle of the island. Day and night, dredges moved back and forth between Galveston Harbor and this canal, dredging up fill from the harbor bottom and spewing it out on either side of the canal in a slurry of water and sand. The first area to be filled was along the seawall, then the rest of the city was filled, section by section.

The lifting operation was one of sheer brawn. Laborers ran beams under the buildings and mounted them on screwjacks that burly men turned by hand. In this way, 2,156 buildings[6] were laboriously hoisted, a quarter of an inch at a turn, until they reached the requisite height and new foundations could be built beneath them. Meanwhile, children climbed rickety catwalks to reach their schools; housewives hung their laundry from lines strung fifteen feet above the ground.

Even substantial structures took to the air. At St. Patrick's Church, a three-hundred-ton brick structure, services continued as it rose to the grunts of laborers manning two hundred screwjacks beneath it. The building was raised five feet. The owners of several elegant Victorian mansions

Before the fill began, engineers traveled the streets of Galveston, marking telephone poles to show how high owners should raise their buildings. Householders jacked their homes up on stilts; sand was poured in around them. In both photographs the man points to the same white line. *(Rosenberg Collection)*

declined to subject them to the rigors of the screwjack. Instead, they let the pumped sand fill their first floor reception rooms or turned them into basements. The lawn of one graceful brick house, once surrounded by a ten-foot wrought-iron fence, is now edged by ornamental ironwork about a foot high — the top of the fence is peeking up through the surface of the fill that now surrounds it.

Though a few wealthy citizens raised their trees as well as their houses, most of the vegetation that had survived the storm was ruthlessly drowned in the slurry of sand. Ardent gardeners boxed their oleanders and stored them on rooftops until they could replant them on higher ground, in top-soil imported from the mainland.

When the project was finished, the city sloped from an elevation of about eight feet on the bay side to as much as twenty-two feet adjoining the new seawall. Water and sewer pipes, gas lines, and trolley tracks had also been raised. To cap the project, a protective layer of boulders, called rip-rap, was laid along the beach at the foot of the wall and a wide brick roadway was built along it, so all Galveston could enjoy the new vista by car or on foot. Although smaller grade-raising operations continued as late as 1928, the initial project was complete by February 1911. The wall itself cost about $1,250,000 and the fill more than $2 million.[7] More than fifteen million cubic yards of sand had been pumped into the city.[8] Property owners paid the cost of raising their own structures.

As a flood protection measure, the wall was a fantastic success. At a ceremony in 1904 dedicating a monument to the effort, J. M. O'Rourke, one of the city's leaders, said it was unnecessary for him to proclaim the wall's merits. "If it ever has an opportunity you will find it well able to talk for itself," he said.[9] On August 17, 1915, he was proved right. Another hurricane made a direct hit at Galveston. Though storm waves scoured its base and carried some of the rip-rap boulders over its top, the wall held. Most of the city remained dry and fewer than a dozen people lost their lives.[10] Advocates and builders of the wall read this relatively low toll as proof that they had challenged the sea and won a bright future for their city.

But Galveston had made a Faustian bargain, and it would pay the price. Bad weather, bad luck, bad timing, and the decision to bet everything on a seawall had put the city on a long, downhill slide.

Beaches and seawalls cannot coexist for long, especially in erosion-prone areas like Galveston. The reason is as simple as it is inexorable: an eroding shoreline is dynamic, but a wall is fixed. The water moves in, the wall stays put. Result: a narrower and narrower beach. Finally, the beach is gone, drowned in a process geologists call "passive erosion." Unless it is constantly extended, raised, rebuilt, and reinforced, no wall is a match for the ocean on an eroding beach. Eventually, it will be undermined and it will collapse. It may even accelerate its own destruction by inhibiting the natural ways beaches respond to bad weather.

Even the mansions of Galveston's rich rose to escape the flood of sand. *(Rosenberg Collection)*

When a beach is threatened by a storm, it rearranges itself to cope. Harsh storm winds quickly carry lighter sand particles on the surface of the beach to the dunes, where the beach has already established reserves of sand. The heavier particles left behind form a kind of protective covering of coarse grains too heavy for the wind to pick up. If waves do bite into the dunes, the sand they carry away collects in underwater sandbars. These are exactly what the beach needs to break the waves offshore and weaken them before they hit the beach itself. The reserve battalions of sand are turned into frontline troops. Eventually, the storm passes. Now gentle swells pick up sand from the offshore bars, carry it inland and return it to the beach.

This system offers every advantage. It operates automatically, requires no government funding, and provides, as a fringe benefit, the fun and beauty of the beach itself. But it has one giant drawback: it only works when people keep their houses, hotels, boardwalks, parking lots, roads, sewer lines, and the like out of the way—and the beach is free to move as it must to respond to storms. Nowadays, however, this kind of infrastructure is all over the coast. It is too valuable to lose, and the cry goes up: build a wall to protect it. Few people stop to calculate that the infrastructure's value derives in large part from the beach—the beach the wall will inevitably destroy.

Sometimes the loss of beach begins at once, with the very placement of the wall. Property owners or civic leaders, eager to preserve as much real estate as possible, position the wall as close to the ocean as they can—in the dunes or even closer. The wall impounds all of the sand behind it. Now, when storm waves attack the beach, the beach does not have access to its reserves of sand. Bars still form, but at a greater cost: a gentle beach slope may turn into a steep decline, increasing the effects of the waves and leaving the beach even more vulnerable. And when gentle swells try to return sand to the beach, the wall stands in the way. The beach's primary defense mechanism for changing sea levels or wave conditions, the exchange of sand between dune and surfzone, is ended.

Other problems also plague seawalls. Even the longest wall must end somewhere, and at that point its presence can cause severe erosion as storms waves cut around behind it. Plus, adjacent beaches are deprived of any of the impounded sand that might once have come into the water and drifted their way. As a result, erosion is often exaggerated at the ends of a seawall.

Whether they know it or not, people who build seawalls commit themselves to the loss of their beach. Once the beach is gone, they are in real trouble. Maintaining even a modest seawall in the face of ocean on-slaughts can be extremely expensive. In extreme cases, the costs can out-weigh the value of the property the wall protects.

In January 1901, while Galveston was still reeling from the great storm, geologists working just inland discovered the Beaumont (Spindletop) oil field. The find brought vast new business to Texas, but wounded Galveston could not handle it. The small settlement further up Galves-ton Bay rose to the occasion; at first a municipal rival, Houston soon eclipsed Galveston, winning its business and then some. While Galve-ston barely held its own, Houston grew into the fourth largest city in the United States.

Galveston's leaders turned to their beach as a source of revenue. With the end of World War I, they began touting the city as a vacation resort. But the beach was beginning to disappear. Within twenty years, the city had lost one hundred yards of sand. People who once watched auto rac-ing on a wide beach were left with a narrow strip of sand at low tide and a gloomy vista of waves on rocks when the tide was high.

A number of explanations were put forward to explain the loss of the beach; a jetty built at the east end of Galveston Island was one plausible villain. But underlying all of these explanations was one simple fact: in building the wall, Galveston had drawn a fixed line on a landscape of un-stable sand. It had protected its homes and offices, its banks and ware-houses. But it had lost its beach and its resort business with it. Prohibition, the Depression, the growing influence of Houston, and—finally—World War II left Galveston to gamblers, prostitutes, and off-duty servicemen. Its own people called it "an open city."

Throughout, the sea continued its attack. Ocean waves began under-mining the rock rip-rap at the base of the wall and flanking its west end. An-other layer of rip-rap was added and, in another effort to protect the wall, thirteen rock ribs or groins were extended like fingers into the surf, to trap sand. These efforts helped, but only a bit. Twice the wall was extended. Today, it is more than six miles long, one of the longest in the world.

But the Gulf is as fast as the builders. The last extension was designed with a ramp at one end, running down to the beach. That ramp is all but useless now. The beach it led to has vanished under water.

There is no way to know if Galveston's leaders would have chosen an-other course in 1900, had they known then what the wall would do to their city. Nowadays, many coastal engineers say, the evil effects of seawalls are so well established that no one builds them anymore. Indeed, some coastal states have enacted regulations banning them.

These regulations stand unchallenged—but only until erosion threatens to send roads or buildings tumbling into the sea. When a breach in a barrier spit left bayside homes in Chatham, Massachusetts, vulnerable to attack from Atlantic waves in 1987, the state's antiseawall provisions were attacked as well. Regulators in North Carolina, one of the first states to limit coastal armoring, look the other way when sandbags are used as de facto sea walls to protect a vulnerable coast road or a resort development. The South Carolina legislature weakened its beachfront building restrictions after Hurricane Hugo generated a storm of litigation from aggrieved owners.

In a contest between the somewhat abstract idea that a wall may eventually damage the public's beach and property owners' all-too-certain knowledge that their buildings are about to fall to the sea, political reality usually dictates the decision: save the buildings, not the beach.

In the aftermath of the 1900 storm, civic leaders in Galveston tried to wipe out any evidence that the ocean once flowed through the downtown streets. People who had placed markers to show how high the water had been were emphatically urged to remove them on the grounds that memories of the storm could only be bad for business.

Today's municipal strategy is different. Once again, Galveston is marketing itself as a seaside resort. But, ironically, the storm that almost killed it is now one of its major attractions. Like other calamities—the earthquake in San Francisco or the attack on Pearl Harbor in Hawaii—the story of the flood draws visitors to museums, shops, and commemorative exhibits in Galveston's restored Victorian business district.

But the city needs more than memories of disaster to thrive as a coastal resort. So Galveston's leaders want to re-create the beach the wall destroyed. Once again, they talk about dredges and pipes and sand slurrys—and money. They want to pump millions of cubic yards of sand into the water in front of the wall, to re-create the city's lost beach. This kind of beach nourishment has become the remedy of choice all over the coast for communities that have armored their beaches out of existence or whose beaches are threatened by erosion from other causes.

Until recently, the federal government, through the Army Corps of Engineers, might have paid much of the bill for such a project. In today's leaner times, though, the federal government is not eager to pay to construct recreational beaches. Its emphasis is on protection from storms. But as Mr. O'Rourke predicted in 1904, the wall has declared itself on this point, and the Corps of Engineers has agreed: as long as the wall stands, Galveston does not need additional storm protection. So the city must find the money itself.

Improvements in meteorology and hurricane tracking make it highly unlikely that a powerful storm could strike the coast of the United States with as little warning as the residents of Galveston had on September 8, 1900. On the other hand, the nation's vulnerable coastal population has increased one hundred-fold since then, so a warning that might have sufficed in 1900 would no longer give today's coastal residents enough time to escape.

Until this century, few people lived near the beach. It was just too dangerous. If they settled along the coast, they built on high areas, well away from the water. Even Galveston's business district and haughtiest residential streets were well away from the Gulf. The hardy few who did build near the beach lived in modest structures they could either move or lose with relative equanimity.

Today, the opposite pattern is well established. Almost half of all construction in the United States in the 1970s and 1980s took place in coastal areas,[11] and demographers estimate that by the year 2000, 80 percent of Americans will live within an hour's drive of the coast.[12] By 2010, the National Oceanographic and Atmospheric Administration says, population density along ocean coasts will be almost four hundred people per square mile, as against less than one hundred per square mile for the rest of the nation.[13]

But the coast is not a stable landscape. Inlets open, heal, and reform. Seaside cliffs erode and slump. Sand shifts. Though we think of the land as terra firma, when we go to the beach on a stormy day we can watch geological change occur almost before our eyes.

Much of this development began before science was able to say precisely what was happening in the geology of the coast. Even today much remains unknown about what happens in the mysterious region where air, water, and land meet. For one thing, research can be difficult to conduct; studying the surfzone is notoriously labor-intensive, unpleasant, and dangerous. Even worse, the nation's increasing commitment to living on the beach has created a powerful force against the application of knowledge already in hand. There is a kind of constituency of ignorance, people who have so much invested in coastal real estate that they do not want to hear how vulnerable it is.

The biggest burst of development on the coast has occurred since about 1970, during a period in which there have been unusually few coastal storms, particularly hurricanes. If weather patterns were now to return to those of earlier years in the twentieth century, the property damage and loss of life could be devastating. On top of that, sea level is rising; if the earth is warming, and mainstream climatologists believe it is, the situation will only get worse. Geologists say 70 percent of the saltwater coastline of the lower forty-eight states is eroding, and some put the figure at closer to 90 percent. The Galveston solution—armoring—can hold off the sea, but only for a while. The best shore protection is a wide, healthy beach. No amount of rock or concrete can make a beach wider. And Nature always bats last at the coast.

CHAPTER TWO

The Great Beach

> . . . I have seen the hungry ocean gain
> Advantage on the kingdom of the shore. . . .
> —Shakespeare, *Sonnet 64*

One June morning in 1992, Steve Leatherman[1] returned to the Atlantic Ocean beaches of Massachusetts, where he had begun his career as a coastal geologist more than a decade before. At that time, in the summer of 1979, he was finishing his graduate work at the University of Virginia and working at the Cape Cod National Seashore, helping to track erosion of the high sand cliffs that run for miles along the Atlantic. Now, he was Dr. Stephen P. Leatherman, director of the Laboratory for Coastal Research at the University of Maryland, and he had joined other prominent coastal geologists and engineers to confront a crisis: erosion was eating away at the nation's coastline, and almost everything being done to combat it was pointless, pernicious, or unaffordable.

The National Academy of Sciences had convened the experts to figure out how to convey that message to the local, state, and federal officials who make policy for the coast. They met at a conference center the academy operates at an old estate in Woods Hole, a rock-bound town on the Cape's southern shore. As they debated the problem and broke into groups to discuss possible remedies, the scientists could look out the windows of their shingled conference building over a lush green lawn running down to the rocks and a tiny beach at the edge of the harbor. Though it was almost summer, the water was still cold, and fog shrouded the meeting. Inside, though, the discussion was hot.

Worldwide, Leatherman was reminding the group, 70 percent of sandy beaches are eroding, and in the United States the figure was at least that high and maybe higher. Rising sea levels could only make matters worse.

But development was continuing in this endangered landscape. How, the scientists asked themselves, could they convince local, state, and federal policy makers to take erosion into account when they planned development? In other words, how could they keep the builders off the beach?

The question was not new. Eleven years earlier, a small group of coastal geologists had confronted it at a conference at the Skidaway Institute of Oceanography, in Savannah, Georgia. They summed up their work in a twelve-page report whose conclusions were stark:

1. People are directly responsible for the "erosion problem" by constructing buildings near the beach. For practical purposes, there is no erosion problem where there are no buildings or farms.
2. Fixed shoreline structures (breakwaters, groins, seawalls, etc.) can be successful in prolonging the life of beach buildings. However, they almost always accelerate the natural rate of beach erosion . . . in the immediate vicinity of structures or . . . along adjacent shorelines sometimes miles away.
3. Most shoreline stabilization projects protect property, not beaches. The protected property belongs to a few individuals relative to the number of Americans who use beaches. If left alone, beaches will always be present, even if they are moving landward.
4. The cost of saving beach property by stabilization is very high. Often it is greater than the value of the property to be saved especially if long range costs are considered.
5. Shoreline stabilization in the long run (10 to 100 years) usually results in severe degradation or total loss of a valuable natural resource, the open ocean beach.
6. Historical data show that shoreline stabilization is irreversible. Once a beach has been stabilized, it will almost always remain in a stabilized state at increasing cost to the taxpayer.[2]

The organizers of that conference, Professor Orrin H. Pilkey and Professor James D. Howard, geologists at Duke University, circulated the report to other geologists and in January 1982 submitted it to President Reagan. "We are a group of concerned coastal and nearshore geologists from coastal areas in the United States," they wrote him. "We wish to call to the attention of your administration our belief that a new approach to management of the American shoreline is urgently needed." The report was signed by eighty-five experts on coastal geology from all over the United States.

But their effort brought no major change in federal policies. In the ensuing decade, the problems of coastal erosion only worsened as development accelerated at the beach. Many scientists at Woods Hole said the meeting itself testified to their failure to educate the people who make coastal policy. "It seems to take natural disasters to focus people's attention on the coast," Leatherman said gloomily, as he showed the group slides of a motel under construction at Cape Hatteras, North Carolina, yards away from the storm-wrecked remains of another building, there on the eroding beach. "The public is making enormous investments along the shore—a trillion dollar line drawn in the sand."

After several days of this, Leatherman was eager to get back to the field, so when the conference ended he left Woods Hole and headed further out on the Cape, to the place he had worked more than a decade before. Cape Cod is shaped somewhat like a bent arm, and Leatherman was headed to the easternmost portion or forearm, known to New Englanders as the Outer Cape. The Outer Cape runs more than forty miles from its elbow at the town of Chatham to its bent wrist at Race Point, near Provincetown. For much of this length, it presents a cliff of sand and clay to the Atlantic. In places the bluffs are well over one hundred feet high, with sandy beaches at their feet.

Retracing his graduate school path, Leatherman was following again in the footsteps of one of America's first coastal surveyors, Henry L. Marindin, an assistant with the U.S. Coast and Geodetic Survey more than one hundred years ago. Accompanied by two observers, two recorders, and a crew of surf-boatmen, Marindin spent the warm months of 1887, 1888, and 1889 mapping the Atlantic coast of the Outer Cape. He and his assistants conducted elaborate surveys along more than two hundred parallel lines starting well inland and running over the cliffs, dunes, and beaches and out into the ocean. Marindin began his survey at Chatham and worked his way north to Provincetown. At each one of his survey lines along the way, he established "permanent bench-marks . . . at points where they were likely to remain undisturbed."[3] These bench-marks, he wrote in a report to the superintendent of the survey, would "provide a base for future comparisons, which will be of value to geologists and others who study the changes in the coastline."[4]

This was the "vast, wild" coast Henry David Thoreau had visited almost forty years before and described in his book, *Cape Cod*. In the intervening years, the Cape's "golden age" of seafaring prosperity had ended as whaling declined. By the time Marindin set out, much of the Cape was linked by rail to Boston and New York, and it would not be long before it would fulfill Thoreau's prediction: "The time must come when this coast will be a resort."

One thing remained unchanged: the sea was chewing at the cliffs. Using old charts, particularly one made in 1856, Marindin estimated that the cliffs of the Outer Cape were retreating at the rate of 3.2 feet a year.[5] Over the entire length surveyed, Marindin wrote, "We find that 32,233,030 cubic yards of earth and sand have disappeared." If this sediment were applied to the fifty-five acres comprising the grounds of the Capitol in Washington, he said, it would cover the statue atop the dome to a depth of sixty-seven feet.[6]

Marindin sent his report to Washington, where it was filed away and forgotten for almost seventy years. But in the mid-1950s, a team from the Woods Hole Oceanographic Institution visited the Outer Cape to reestablish the survey points and recompute erosion rates. One of the team was Graham Giese, a surveyor newly out of the army. The work was difficult, he wrote later. Most of Marindin's original wooden benchmarks had not survived. Some had apparently washed away. Wind had buried others deep in sand. Still, the team found seventy-four markers and remeasured the distances from marker to bluff-edge. This time, Giese and others on the team calculated erosion rates as averaging 2.5 feet per year.

They replaced the old markers with copper rods more than twenty feet long, driven into the sand. The top four or five feet of each were clad in concrete, with a brass cap mounted on top. Stamped into the cap on the spot were its number, precise measurements of its position, and its mean sea level.

Giese returned to the bluffs in 1979. By now he was one of the leading authorities on the geology of Cape Cod, and it was his group Leatherman joined. Using the notes Giese and his colleagues had made more than twenty years before, they began the search again. But even more of the benchmarks had been lost to erosion or covered by windblown sand, and the group could find only eighteen. Still, they were able to determine that since the survey in the 1950s erosion on the Outer Cape was continuing at a rate of 2.2 feet a year. More must be done to protect the remaining markers, they declared, "in order that this valuable source of information be maintained for future surveys and updating of cliff changes."[7] Resolved to make things easier for the next search, the team marked each monument they found with a second pipe, an eight-foot length of corrosion-resistant iron. Leatherman stood on a stepladder and used a sledgehammer to drive them into the sand. "This is the only place in the United States that has this kind of benchmarks that go back over one hundred years," Leatherman said, as he set out to find them again. "This is unique."

First stop was in Lecount Hollow, at a parking lot on the bluff above a sandy beach in the Cape Cod National Seashore. Though it was a cool, overcast day, a few hardy bathers had set their towels and chairs out on the sand at the bottom of a steep wooden stair. Leatherman, his reddish hair blowing in the wind, turned away from the ocean and, taking his bearings from an old gray house and a twisted pine tree he had described in his field notes almost fifteen years before, stepped slowly across the bluff, his eyes searching the knee-high beach grass, goldenrod, wild roses, and heather. Here and there were bright red flowers, dubbed "British soldiers" for their red color, and here and there were holes in the sand, the entrances to the burrows of ghost crabs.

After only a few minutes, he spotted what he was looking for: a nondescript gray pipe sticking up about a foot from the sand. "This is it, this is the pipe we put in," he said. He knelt and began to brush away the sand near its base. "Wind has carried sand up on the dune," he said. In a moment, he had uncovered the top of a concrete post, four inches square, buried under several inches of sand. Mounted on its top was a round brass plaque, two inches in diameter and somewhat green after years under the sand. This post, called a monument, was marked, "Woods Hole Oceanographic Inst. Coastal Survey—NO [for monument number] 87, AZ [for azimuth] 253.47 and MSL [for distance above mean sea level] 43.1 feet."

When Leatherman found this monument in 1979, the parking lot had not been built yet. And the landmarks Marindin and Giese had noted in previous surveys were almost useless. "The original road Marindin took surveys on is gone—washed away," Leatherman said. "It was used in '57 but it was gone by '79. We had air photos, but they showed roads and telephone lines that didn't exist any more." Instead, he, Giese, and the others used metal detectors. "We spent four days looking here to find this monument," he recalled. "We found every copper penny that ever fell on this dune."

Leatherman made new notes on the pattern of vegetation and building in the area, but here too things had changed. The beach heather community he had noted in 1979 was still thriving, but the twisted pine tree he had used as a landmark was nearly dead. It would probably not be around to guide the next team looking for the monuments. Even more than vegetation, the buildings had changed. There was nothing left of one house he had noted in 1979 except the remains of a concrete septic tank sticking out of the edge of the eroding bluff. The house itself had fallen away.

Later in the day, he found another monument at the edge of what is now the driveway of another cottage. The concrete rose out of the ground at an angle ("someone evidently backed over the monument," Leatherman observed), and its brass plaque had been knocked off. This is why the scientists who use the monuments strive to keep them unobtrusive. "People like to destroy them or take them as souvenirs," Leatherman said. "Idiots." The site of still another monument was identified in his old notebook as lying on a particular street, "44 feet NNE of NE corner of gray shingled cottage with green shutters." But now there were six gray cottages with green shutters on the old dirt road.

At still another monument, the point of reference was a gray shingled bungalow with white trim, perched about fifteen feet from the edge of the bluff. This house was still a useful landmark, alone on the bluff. But the beach is retreating here about three feet a year, Leatherman estimated. "That house has a five- to ten-year tenure before it's going to go over the edge," he said.

At the end of the day he found one last benchmark, between two telephone poles along a road that had not yet been built when he found it the first time. "I can remember standing here and not wanting to leave," he said, recalling 1979. "I was so proud to find it. Everything was measured from a road that was out in the ocean."

The ocean laps at the land in many a rhythmic cadence. The most familiar is the surf, whose lulling rush makes a day at the beach so relaxing. But other oceanic rhythms shape the coast. There are the tides, the seasons, and, underlying it all, the rise or fall of the sea. The tides ebb and flood with the varying gravitational pull of the Moon and Sun as Earth follows its yearly path through the heavens. The tidal range—the difference in water levels at low tide and high tide—varies from place to place along the coast, as geography varies. And everywhere the tidal range has its own rhythms, narrowing with neap tides when the moon is at its first or third quarter and there is relatively little difference between high and low tides, and widening again two weeks later, when the moon and sun line up and the pull of gravity is felt the most, producing the extreme high and low water of spring tides. When the moon is nearest the earth, at its annual perigee, its gravitational effects are exaggerated even more, in what is called a perigean tide. The tidal range is greatest in perigean spring tides.

The seasons add another rhythm to the coast, as gentle summer weather brings long, slow swells that tend to carry sand up on the beach, followed by winter storm waves that carry sand away. In most places, the beach narrows every winter, only to start widening again in spring.

Swamping all of these comings and goings is the rise and fall of sea level that occurs over tens of thousands of years as the earth warms and cools. When the earth is warm the seas are high. In fact, the seas once reached well into the United States, gnawing away at the land far into the Appalachian plateau. Richmond, Virginia, Raleigh, North Carolina, and other modern cities are, in their way, evidence of this geological past. They were built to capitalize on waterpower at the fall line, the boundary where the rivers of the plateau fall away to the broad Atlantic coastal plain, washed almost flat by the wave action of ancient times. On the plain itself, limestone and phosphate quarries profit from the remains of millions of tiny sea creatures who once lived there, under seawater.[8] When the earth cools, however, and more of the earth's water is frozen up in spreading glaciers and ice sheets, the sea falls away. Ice and sea have been dancing to this music for millennia.

Cape Cod is the product of the the last advance of the ice, an Ice Age that began about seventy thousand years ago, when great ice sheets began to spread south from Canada, scraping every rock and pebble and bit of soil before them until the glacial lobes reached what are now Long Island in New York and the islands of Martha's Vineyard and Nantucket, off the south coast of Massachusetts. Sea level was about three hundred feet lower then, and the place where the coastline ran in those days is today as much as seventy-five miles at sea.

Starting about eighteen thousand years ago, however, the earth began to warm and the glaciers began to melt and retreat; their retreat accelerated about twelve thousand years ago. As they melted, they left the soil and rock they had carried behind them. Meanwhile, water from the melting ice was flowing into the sea, which began to rise, flooding the low Atlantic coastal plain and moving inland as much as one hundred feet a year. By seven thousand years ago, rising water had made an island out of Georges Bank, a 150-mile ridge northeast of Nantucket. By four thousand years ago, when the earth's climate more or less stabilized, the banks were under water and Nantucket and Martha's Vineyard had become islands off the coast of what we know today as Cape Cod.

Since then, the earth's climate has been more or less stable, and the sea's rise has slowed so much that people came to assume that the boundary between ocean and land was more or less permanent. But it is not. Worldwide, sea level continues a slow rise. And if global warming is the problem most climatologists think it is, the rise will accelerate in the twenty-first century.

But sea level rise alone is not enough to explain the widespread erosion of the United States coast. Among the other factors at play are the movement of the tectonic plates that form the earth's surface, sand supply, and the action of waves, winds, and currents. Differences in these factors explain the different appearance of the East and West coasts of the United States.

The East Coast is the product of the ancient collision of the North American and Atlantic plates, two of the tectonic plates that form the surface of the earth. Among other things, this collision produced the Appalachian Mountains, a relatively old mountain chain whose wearing away over millions of years provided some of the sediment that built the broad continental shelf of the East Coast and the low-relief, sandy islands, backed by bays, lagoons, or sounds, that run along its length from New England to Mexico, the longest string of barrier islands in the world.

A few of these islands are high and wide, with extensive dune fields and well-established maritime forests. Others are narrow with low and irregular dunes and few trees. Still others are little more than strips of sand with poorly developed dunes. When these barrier islands are attacked by rising seas, their natural response is to back out of the way. Storm waves wash over them, carrying sand to the back side in overwash fans of sediment. Marsh plants colonize these sand deposits and trap still more sand. As the islands erode on the ocean side, they grow on the sound side, maintaining their overall shape as they shift inland to higher ground. If the island is frail, storms may break through it, allowing subsequent high tides to carry even more sand through the breach to the back side of the island, where it collects in flood tide deltas.

On the East and Gulf coasts, where the land slopes gently toward the coast, an island can slowly move out of the way of rising seas, if the change in sea level is slow enough. It retains its approximate beach width and configuration, only now it is further inland. Most of the barrier islands have been migrating like this for thousands of years, their movements timed according to the amount of sand available on these islands and the rate of sea level rise.[9]

Onshore winds also help islands move inland by blowing sand from the beach into the dunes and from there further inland. Sand fences that beachfront property owners use to trap sand and build protective dunes in front of their homes take advantage of this process. Even something as small as a sprig of beach grass or a tiny piece of driftwood may be enough of a barrier to trap these grains of sand and form the base of what will grow into a dune. But wind is a minor player in barrier island migration. The two most important processes are overwash and inlet formation, particularly on the Atlantic coast south of Cape Cod.

Things are somewhat different on the West Coast, where the Pacific plate is sliding or subducting under the North American plate. The Rocky Mountains, the coastal mountain ranges, and the narrow continental shelf are among the results of this new and vigorous tectonic collision. At the same time, the plates are sliding past each other. The North American plate is moving northwest, while the Pacific plate moves southeast.

When rising seas advance on this coast, low areas such as river valleys are flooded. The hills and mountains between them remain as headlands protruding into the advancing sea, their beaches fed by sediment washed off inland hillsides and carried downriver. When these headlands feel the full brunt of the ocean waves, they begin to erode and collapse, providing new rock, gravel, and soil to be ground down by waves. In this way, these beaches, too, maintain their form, even as they move inland. The continuing violence of the tectonic collisions on the West Coast is pushing the land up, so there is somewhat less erosion caused directly by sea level rise. On the other hand, water depth drops off dramatically close to shore, and beaches hug the shoreline. Although many stretches of the West Coast are lined with barrier spits that resemble East Coast beaches, the typical West Coast beach is a narrow strip of sand backed by steep cliffs.

Sea level determines where the waves will break. But, to a great degree, the sun determines how they break, because it is the sun that warms the air to generate the winds that power them. In effect, breaking waves transfer solar energy to the coastline. The size of the waves depends on how fast, how long, and over what distance — or fetch — the wind has blown. The harder, longer, and farther the wind blows, the bigger the waves.

The closer they are to the wind that made them, the choppier waves are. As they spread out and calm down, they become more regular and are known as swells. When waves move into shallow water, water at the surface moves faster than water caught up in the friction of the bottom. The waves steepen and become unstable. When the water is only a little deeper than the height of the wave, the wave breaks, releasing much of its energy to reshape the beach.

Meanwhile, water moving over sediment-like sand sets its grains in motion, picking some of them up and carrying them along. Some of this sand moves along the bed in ripples that form in mysterious ways in the oscillation of water on the bottom. But much of it moves in the water itself. The faster the water moves, the more sediment it can carry. When the water slows, the sediment drops out, or precipitates. This process helps explain how breaking waves can pick up sand in the turbulence of the surfzone and drop it on the beach when they run out of steam. It also explains how storm waves can have so much energy that they bounce off a beach, carrying sand away with them only to drop it in deeper water offshore. In deep water, sediment at the bottom rarely feels the movement of the waves at the surface. It takes a big wave to move sediment under thirty feet of water.

If waves broke exactly parallel to the shoreline, the life of a sand grain would be a simple one. It would come onshore in mild weather and go off again in storms. But waves do not come in on the square. Depending on many factors, notably the predominant direction and strength of the wind,

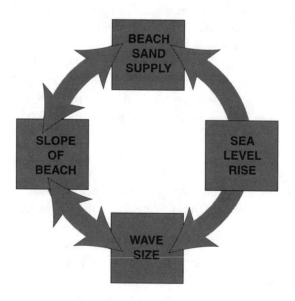

The shoreline is shaped by the interplay of sand, sea level rise, topography, and weather.
(*Jen Christiansen*)

waves hit land at a slight angle, often so slight as to be invisible to sun-bathers on the beach. But it is enough to set up alongshore (or longshore) currents that can move a veritable river of sand in one direction or another as winds shift along the coast, especially in storms.

The sum of all of this movement of sand, usually measured over a year, is the beach's net littoral drift, and in some places it is enormous. On the Outer Banks of North Carolina, where waves are as energetic as they are anywhere on the East Coast, geologists estimate that as much as seven hundred thousand cubic yards of sand may move from north to south along a given stretch of beach each year. (A good-sized dumptruck holds ten cubic yards or more.) On the far reaches of Cape Cod, a heavy flow of sand from south to north has built miles of new beach, called Province-lands, where the Cape's forearm bends at the wrist. In other places, shifting winds may produce a net drift so small that a single episode of unusually severe weather can alter the beach for decades.

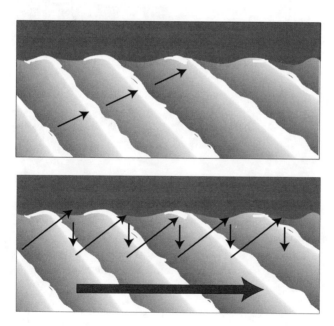

Because they almost always reach shore at a slight angle,
waves generate currents that flow along the shores. These
currents carry sand along with them in littoral drift.
(Jen Christiansen)

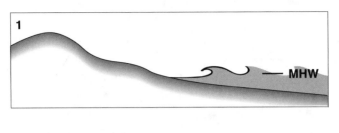

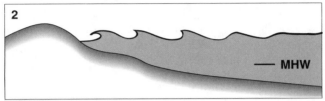

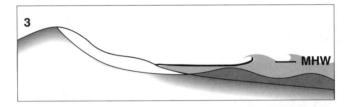

In good weather (1), waves break as the nearshore slopes up to meet the beach. In heavy weather (2), higher waves attack the dunes, carrying sand away offshore (3), where it collects in one or more bars (4). When calm weather returns, gentle swells will carry some of the sand from the bar back to the beach. *(Jen Christiansen)*

When big storms strike, the beach changes before our eyes. Water levels rise and waves come quicker and bigger. They break further inland, chopping away at the beach face. If the storm goes on long enough or the water is high enough, the waves may chew away at the dunes or cliff behind the beach and carry sand away. But the sand is not actually lost. In a process still not fully understood, the surf transmutes this sand into bars that, in turn, break the stormwaves' force before they reach the beach. In a big storm one, two, or even three sandbars may form off the beach, each one dissipating the waves' energy so that when they hit the beach they run up, sink into the porous sand, and ooze back out, taking little sand with them.

Meanwhile, as wind and wave winnow away the lightweight particles and blow them inland, the remaining heavier bits of broken shell, rock fragments, and grains of dark red garnet or other minerals act as armor for what lies beneath. (After a storm, these minerals form a dark layer on the beach. When normal weather returns, this dark layer will be covered by a layer of finer, lighter grains, enabling an observant person to track the beach's recovery from the storm by its changing color, and its long-term storm history by slicing into it and "reading" the layers of color under its surface.)

By the time the storm ends, the beach has been transformed. If it had dunes, the dunes may be gone. If it was backed by cliffs, the high tide may be lapping at the foot of cliffs and there may be no beach at all. The beach appears to have eroded—and it has. But in another sense it has not. It has simply used its sand resources in its own defense. The sand is in the water now but, with luck, good weather, and gentle swells, the sand in the bars will begin to migrate back to the beach. It may take weeks, months, or even years but—if nothing interferes—eventually the beach will regain its shape, more or less.

Erosion occurs when more sand moves out of an area than moves into it. This idea is so simple, so obvious, that is a shock to discover that coastal engineers, planners, and developers have only recently begun to act on it.

Douglas Inman, director of the Center for Coastal Studies at the Scripps Institution of Oceanography and another participant at the Woods Hole meeting, was one of the scientists who developed the intellectual framework for this concept in the 1970s. He theorized that beaches are the product of three factors: a supply or source of sand; something to move it around—a transport mechanism; and its final destination, or sink. The

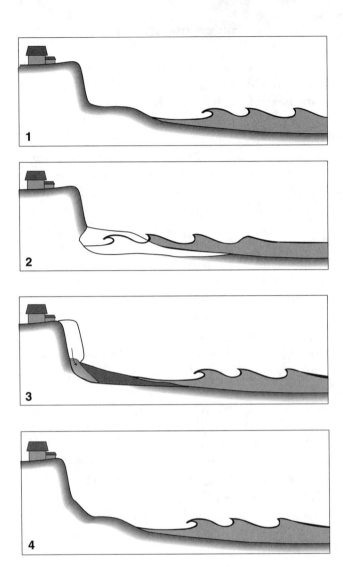

On a cliff coast, beaches are usually narrow (1). When storms strike, waves eat away at the base of the bluff (2). Eventually, the bluff collapses (3), providing more sand for the beach (4). *(Jen Christiansen)*

three together form what Inman called a "littoral cell." As he saw it, the coast as a whole is as a series of these cells.

"The concept that I devised of the littoral cell—the source, the transport of sand, and the sink of sediment—is a very fundamental but important concept," he said years later. "Because if you know the source and the transport path and the sink, then you can predict what will happen when you interrupt this."

Though crucial details of how sand moves remain unknown ("People have no concept of what moves sediment along the coast," Inman says), this theory fits so many places so beautifully that it is a mystery even to him why it took so long for someone to articulate it. "This was not known when these were built in the thirties," he said, referring to jetties, breakwaters, and other pieces of shoreline engineering constructed on the southern California coast. "It's not clear to me why it wasn't known, but it wasn't."

In California, Inman's original study area, cells are stretches of coast separated by rocky headlands. In each cell, the natural source of most beach sand is water flowing down the mountains in rivers and creeks. As this water erodes stream and river beds, it picks up sediment and carries it along, only to drop it when it reaches the ocean. There, the sediment is distributed by coastal currents. As the currents carry it along, the sand eventually falls into one of the ancient river valleys, or submarine canyons, that approach very close to the shoreline. Each of these canyons is a sand sink. Water moving over it slows, and the sediment sinks into the canyon, where it is lost forever.

This loss of sand has little effect on the beach, as long as fresh supplies of sediment continue to wash down from inland hills. But in the last fifty years or so, most of the rivers and streams running into the Pacific have been lined with concrete for flood control or blocked outright by flood control or power projects. Less sediment gets into the water, and much of that ends up trapped behind dams. Starved for sand, the beaches look for a new source of sediment and turn to the cliffs that line most of the beaches of the West Coast. Waves begin gnawing away at the bluff base, the bluff fails, and a fresh supply of sediment collapses onto the beach. Evidence for this process is abundant. It can be seen in the century-old maps of San Diego showing streets that no longer exist and the offshore rock formations, called "stacks," that stand alone, the bluffs that once surrounded them long gone.

Cliff erosion can be highly episodic. A stretch of coast may survive unscathed for years in relatively mild weather, only to retreat dramatically in

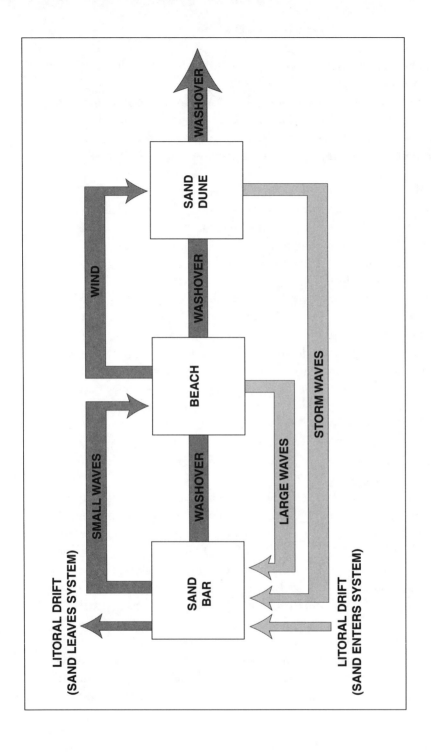

Schematic shows the forces that carry sand on and off the beach. *(Jen Christiansen)*

the face of one severe storm. For example, one stretch of cliffs at Santa Cruz, California eroded about twenty-five feet in the interval from 1931 to 1982, or at an average rate of about six inches per year. But one fierce storm, in January 1983, removed about forty-six feet of bluff top. This single episode of bad weather almost tripled the long-term erosion rate from six inches per year to sixteen inches per year. As two California researchers pointed out at the time, "These kinds of observations argue for using average erosion rates with great caution and making every effort to research the historic changes as far back in time as possible."[10] Even water from gardening and septic draining can cause cliffs to give way, by weakening their integrity so that they collapse more readily under the assault of the waves.

The same sort of source-transport-sink system exists on the East Coast, though its operation is not quite so obvious. Here, the rivers that carry sed-

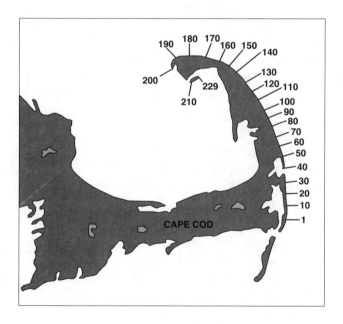

Henry Marindin, a geologist with the U.S. Coast and Geodetic Survey, established more than 200 benchmarks along Cape Cod. Many of them have eroded away since Marindin roamed the Cape 150 years ago.
(Stephen P. Leatherman)

iment to the coast usually dump it in estuaries behind barrier islands, and much of it never makes it into the beach system. But beaches are also fed from eroding headlands. On the Outer Cape, for example, erosion of the cliffs in the middle of the forearm feeds beaches both to the north and to the south. "Here in southeast Massachusetts the sediment which makes up the beaches and the bars, barrier beaches, and dunes is almost entirely derived from erosion of glacial deposits," Giese said. "If you prevented the cliffs from eroding, the beaches would be gone in about eighty years."

As on the West Coast, longshore currents are the transport mechanism that carries sand along the barrier islands until it reaches an inlet. If the inlet is in its natural state, the sand will be caught up in its tidal currents and precipitate inside and outside its mouth, in flood or ebb tide deltas. If the inlet is kept navigable by jetties, sand will pile up behind them. If the inlet is dredged, the sand is typically dumped in deep water offshore, in which case it will probably never make its way back to the beach. One way or another, inlets are sediment sinks.

Though the idea of littoral cells is relatively new, the idea that the coast is mutable has been obvious for a long time to anyone who wanted to see it. It has been unusually clear in Chatham, Massachusetts, the town founded in the seventeenth century at the elbow of Cape Cod, and Marindin's starting point two hundred years later. A pattern of coastal change has repeated itself several times since Europeans first established a harbor there at Pleasant Bay, a shallow anchorage protected by a barrier spit. This spit, in turn, is fed by sediment carried in littoral drift from the Outer Cape's eroding cliffs to the north.

The entrance to the harbor is a small inlet that migrates southward and shrinks as the spit grows. Eventually, the inlet becomes so narrow that if storm waves wash over the barrier and into the harbor, the inlet cannot accommodate the water on its way back to sea. So the water feels along the back of the spit for a weak spot, where it bursts through and flows quickly out to sea, widening the breach as it goes.

When calm weather returns, currents continue to carry sand south. But now it is trapped in the ebb tide and flood tide deltas of the new inlet. South of the new inlet, the spit is starved of sand and starts to break up. Eventually, the spit begins to grow again from the north, fed by infusions of sand in the littoral drift—perhaps as much as three hundred thousand cubic yards a year. Once again, the spit starts pushing its way south until, after another one hundred and forty years or so, the landscape has regained its old form, ready for another breach. Until the spit reforms, though, the town of Chatham is deprived of its protection against North Atlantic waves.

When Samuel de Champlain explored the coast in the early seventeenth century, the area was recovering from a breach and the barrier was growing south. It breached again in 1840, and again the spit recovered. By the late 1970s, it had grown so far south that everyone in Chatham was anxiously awaiting the next breach. The Chatham Conservation Commission hired Giese to study the situation and make recommendations about how to deal with it. But there was little he could do other than describe what lay in store for the town. Every winter, when northeasters pummeled the coast, Chatham held its breath.

January 2, 1987, dawned to a perigean spring tide and a fierce snowstorm. The wind from the northeast was blowing fifty miles an hour, and waves more than ten feet high were breaking on the spit. On the landward side of Pleasant Bay, townspeople gathered at the Chatham Lighthouse to see if this was the day their protective barrier would give way. At high tide, shortly after noon, the ocean surged across the barrier and into the harbor. A few hours later, when the tide fell, this water broke through the spit and rushed back out to the sea. By the next day, the breach was well established. A new channel was running between the inner harbor and the ocean, widening with every tide. In the weeks that followed, the barrier south of the breach eroded furiously, widening the new inlet still more. Soon Chatham itself was under assault from the Atlantic.

Like the other two breaks that have occurred since the European settlement of Cape Cod, this breach was directly east of the bluff occupied by the Chatham Lighthouse, the third unlucky structure built on the spot since President Thomas Jefferson appointed the first keeper in 1808.[11] The first lighthouse, built of wood, was condemned in 1841. The second, a sturdier brick structure with twin towers, was built the same year, further inland. But when the barrier breached in 1847, Atlantic waves began to batter the previously protected town streets. In 1870, the twin towers of the lighthouse were 228 feet from the edge of the bluff. By 1874, only 190 feet separated them from the water; two years later half of that land was gone and in 1879 the south tower slid over the bank. The other followed fifteen months later.[12] The current lighthouse was constructed still further inland.

Though Giese's report to the Chatham Conservation Commission had offered no help in preventing damage, it did have one important effect:

"Because the community generally accepted the inevitability of barrier breaching and the shoreline changes that would accompany it, the Conservation Commission was able to enforce more stringent restrictions on coastal development than would have been possible otherwise."[13] Still, there have been fierce arguments, even lawsuits, from property owners challenging state restrictions on walls and other shoreline armor that they were desperate to install to protect their homes.

After the 1847 breach, a row of houses fell to the surf. Others were moved to higher ground, and old-timers in Chatham say that many of the picturesque clapboard cottages along the town's main streets once occupied more precarious perches on the eroding bay.[14] More than a century later, moving was not an option. Houses were no longer modest and there was no longer any place to move them. Since the breach, Chatham has lost bayside houses worth millions of dollars, and now the lighthouse itself is threatened yet again. "We are looking at quite a long period of vulnerability," Giese says.

When glaciers covered Cape Cod, and the Atlantic shoreline was far to the east of where it is now, average global temperatures were only about nine degrees colder, on average, than they are today. Today, global temperatures are on the rise and, possibly as a result, the seas appear to be rising too, relative to land. The Intergovernment Panel on Climate Change, a United Nations organization, anticipates a 0.33 to 1.1 meter (one to three feet or a bit more) rise in sea level by the middle of the next century. "That is not a worst case," Leatherman told the Woods Hole gathering. "It's what is called the design case." In the Netherlands, "where they take this really seriously," the authorities are planning now for a rise of more than three feet, he said.

The effects of such a rise in relative sea level would vary widely from place to place, depending on local geology. In areas of the East Coast, where the coastal plain rises ever so gently from the shoreline to the fall line, the rule of thumb is a one-foot rise in sea level sends the shoreline back one hundred to two hundred feet. "That's the thing people are really going to have to start thinking about," Leatherman said. Already, Inman told the Woods Hole gathering, measurements show that sea level is rising thirty to forty centimeters (about seven to ten inches) every century.

On the Gulf Coast, particularly the coast of Louisiana, compaction of river sediments, marsh damage, withdrawal of hydrocarbons from deep in the earth, and a host of other human interferences have already created catastrophic relative sea level rise. Geologists there estimate that in a matter of decades some of the state's coastal parishes will have virtually washed away.

The situation is somewhat different on the West Coast, where the collision of tectonic plates is causing much of the land mass to rise—though still not quickly enough to outpace the rise in sea level. Sea level there is rising at about half the rate of the East Coast. "We're on a collision coast," Inman told the Woods Hole group. "That's why we have such a narrow shelf, sea cliffs behind, volcanic activity, faulting—all of this goes with this kind of coast."

By contrast, he said, the East Coast "is a low coast line, it's a trailing edge coast and the effect of sea level rise is to let the barrier migrate inland and so you have something like, on the Outer Banks of North Carolina, average landward migration of the order of five to six feet per year. And that means that over a ten-year period you're talking fifty to sixty feet. This is appreciable.

"On the West coast we have an entirely different situation with half the rise. Our beaches are contained against a shelf that's cut into the coastline, notched, and so the beaches can't move at that rate—they can't move any faster than the cliff erodes."

Despite rising seas, much of the Outer Cape probably looks as it did a century and a half ago, when Thoreau tramped its cliff edges and marveled at its landscape. But with the bluff edge retreating more than two feet a year, the seaside path he walked is now probably hundreds of feet offshore. Thoreau himself realized what was happening in the geology under his feet. "Anything built on top of the cliff is doomed," he wrote in *Cape Cod*.

At the Woods Hole gathering, Leatherman spoke of the futility of trying to maintain stability in such an unstable situation. "Our system is based on the assumption of static equilibrium," he said. "But there is no stasis at the shoreline. When people are building large buildings on the edge of the shoreline, why are they astounded when the shoreline shifts?"

CHAPTER THREE

Armor

man is never on the square
he uses up the fat and greenery of the earth
each generation wastes a little more
of the future with greed and lust for riches
—Don Marquis, "What the Ants Are Saying"

In 1797, the story goes, President George Washington ordered the construction of a lighthouse on the cliffs that rise sixty feet or more from the beach at Montauk, at the eastern end of Long Island. The sea was eating away at these cliffs, but he estimated that if the lighthouse were built three hundred feet from the bluff it would be secure for about two hundred years. Two centuries later, Washington was proven correct. The beacon still stands, but in a precarious position, only fifty feet from the cliff edge.

The steady destruction of the Montauk cliffs is part of a natural process that began at the close of the last Ice Age, as the glaciers that had spread down from the north began their melting retreat. The boulders, cobbles, gravel, and sand they had scraped down from the north was the raw material that formed Long Island, the long finger of land that runs from Brooklyn and Queens in New York City, 150 miles east to Montauk Point.

As the glaciers melted and the seas rose, Atlantic waves eventually began gnawing at the island's east end. Material that had consolidated into cliffs crumpled under the onslaught, throwing dirt and rock into the sea, where the constant grinding of the surf reduced it to ever finer grains of sand. Currents picked up this sand—as much as two hundred thousand cubic yards a year—and carried it west along the new coast. As sand was removed, the cliff-crumpling continued, creating new sand in its place. By the time Washington was siting the lighthouse, the action of the waves and currents had carried off enough Montauk sediment to form a string of barrier beaches along most of Long Island's south shore.

These beaches were spectacular expanses of fine-grained, light gray sand, hundreds of yards wide, backed by grass-covered dunes as much as fifteen or twenty feet high. Because they faced south, they were shielded from the worst wrath of winter northeasters. The currents constantly carried sand away from these beaches, but replenished it all the while by eroding the East End cliffs. Meanwhile, sand that washed or blew across the barriers in storms became the base for expanding marshes that slowly grew back into Shinnecock, Moriches, and Great South Bays. As rising sea levels sent the shoreline moving slowly inland, these marshes in turn became the base for new dunes.

Two storms in the 1930s changed all that. One, in 1931, cut an inlet across the barrier island about halfway along the South Shore between Montauk and Manhattan. Left to itself, this inlet might have closed naturally, and the natural flow of sediment would have resumed. But the new inlet offered valuable access to the ocean, so dredgers and jetty-builders kept it open. Moriches Inlet became a permanent part of the Long Island landscape, and the westward flow of the river of sand was blocked. Seven years later, the great hurricane of 1938 cut another inlet, between Moriches Inlet and the Montauk cliffs. This breach, named Shinnecock Inlet, was also kept open. Since then, the beaches west of the inlets have been eroding fast.

Meanwhile, more and more people were building on the barrier islands, especially after the end of World War II, when New Yorkers streamed onto Long Island. Now the erosion of the barriers was a crisis. By 1954, the state of New York was looking to Congress and the Army Corps of Engineers for help, and Congress authorized the corps to intervene. The result, in 1960, was a report that recommended several steps to protect coastal real estate in the reach from Shinnecock Inlet west past Moriches Inlet to Fire Island. Since development on Fire Island was still comparatively sparse, much of the project would deal with the area between Moriches and Shinnecock Inlets, an area known as Westhampton Beach. Among other steps, the report recommended: widening the beach by pumping in thirty-four million cubic yards of sand; removing or elevating some eighty buildings; planting dune grass; and, "if and when experience dictates their need,"[1] constructing as many as fifty groins to hold the new sand in place.

Groins, often incorrectly referred to as jetties, are short rock ribs that begin on the beach, under the sand, and run into the sea, sticking out like fingers into the surf to trap sand. In the corps's plan, these groins would preserve beaches by stopping the currents from carrying sand away. Construction would begin at Moriches Inlet, at the downdrift or western end of the project, and move eastward, so the groins would retain as much as possible of the new sand. State officials reviewed this report and told the corps, "We strongly suggest that the number of groins to be authorized should not be less than 50."[2]

But the corps never completed this plan. The first phase of the project began in 1964 but, in a crucial mistake that would haunt the area for decades, the work did not start as planned at the west end, where groins would trap sand that had already passed across the island. Instead, at the insistence of influential shorefront property owners, the project began in the east, with the first eleven groins constructed in the middle of Westhampton Beach.

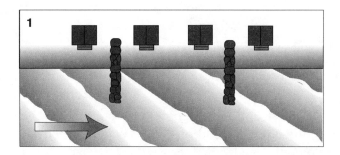

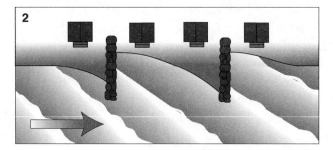

Groins trap sand moving in the littoral drift, helping some beachfront property owners but robbing others of sand.
(Jen Christiansen)

Groins at the east of Westhampton Beach, Long Island (top) trap sand, starving the beach to the west. *(Stephen P. Leatherman)*

As might have been expected, erosion west of the groin field accelerated immediately, provoking loud demands for more groins. Four more were built, but they only shifted the erosion hot spot farther west. And in 1970, prompted by budget problems and complaints from environmentalists who felt that the groins were already doing more harm than good, the government of Suffolk County, where the groins were being built, stopped paying its share of the cost. The corps halted the project. A bad idea, badly executed, turned into a disaster.

Though property owners updrift were sitting pretty, behind a wide expanse of trapped sand, owners of houses to the west of the groin field found themselves in a desperate situation. Currents were still carrying their sand away, but the groins were keeping new sand from moving in to replace it. Their beach was starving. Severe erosion had destroyed much of its dune line and left one row of houses on each side of the two-lane blacktop—called, ironically, Dune Road—that runs the length of the narrow barrier. Still, in spite of what was now an obvious problem, people continued to build large new houses on the ocean side. To qualify for federal flood insurance, they built them atop wooden pilings. At high tide, they looked like absurd water birds, their toes in the surf.

On the bay side of the road, the houses were older, lower to the ground, and smaller. They faced a different threat. As storm waves washed more and more sand over the increasingly fragile island, their owners fought to keep sand out of their yards, their driveways, and even their living rooms.

Meanwhile, the groins themselves were not immune from erosion. The westernmost groin was quickly flanked by waves that ate into the sand-starved beach behind it, ultimately leaving it standing alone, out in the water. It was reinforced with corrugated metal panels hammered into the sand, but waves undercut this sheet-piling, too.

On Halloween 1991, a fierce northeaster struck much of the East Coast, including Long Island. By now the island west of the groins was virtually flat and barely above sea level. The beach had no reserves of sand to throw back at the storm waves, and the water moved across the entire island, pushing beach sand with it and half burying some of the bayside cottages. The storm wrecked many houses and left others leaning crazily on their pilings. (Leatherman had a narrow escape when he visited the battered island to inspect damage there. So much sand had been lost that he was forced to thread his way along Westhampton beach through the pilings holding up the damaged, oceanfront houses. He was walking under one of them when its kitchen floor suddenly gave way, dumping a good-sized

refrigerator at his feet. He was not injured.) Of some houses, nothing remained but concrete septic tanks jutting from the sand. The island had
narrowed until it was only about two hundred feet wide.

Though the beach recovered a bit the following summer, the winter of
1992–3 brought a series of devastating blows. A December storm cut two
breaches through to Moriches Bay. One began healing naturally when
the storm ended, but the larger breach, which came to be known as Little
Pikes Inlet, widened as fierce storms struck again in January and March.
A portion of the island where 250 or more homes once stood was cut off,
reachable only by an amphibious vehicle Suffolk County had acquired for
that purpose from the army. Ninety of the houses were still standing, but
not even their owners were allowed to visit. The place was simply too dangerous. To make matters worse, marauders from the mainland began taking their powerboats across Moriches Bay to the island, where they van-
dalized and robbed the houses that remained.

Homeowners in the affected area had already filed a class-action suit
against Suffolk County and now expanded it to include the state and fed-

Storms in 1991 cut an inlet at Westhampton Beach. Federal, state and local agencies spent millions to repair it. (*The New York Times*)

eral governments. The owners asserted that the erosion control project had destroyed their beach and threatened their homes. To advance their suit, and relying on decades of real estate and voting rights precedent, property owners in the abandoned area incorporated it as the village of West Hampton Dunes. Even people whose land was underwater were permitted to claim the place as their legal residence. Though no one lived there, the new town had an official population of 580.

The suit was settled in October 1994. Under the terms of the settlement, federal, state, and local agencies would cooperate to build a new, four-hundred-foot beach, with 4.5 million cubic yards of pumped-in sand, along with protective dunes as much as fifteen feet high. The two groins closest to the eroded area would be shortened, in hopes that more sand would find its way around them. To forestall questions about why taxpayers should pay for this beach, the property owners agreed to open it to the public. The beach, which would run two miles from the site of the breach to the border of a state park at the east end of the barrier, would be accessible with six walkways from Dune Road. (This was cold comfort to advocates of public access, though, because there is no parking on Dune Road.) People whose houses had been lost would be able to rebuild if, after the work was completed, their lots were at least seventy-five feet deep and their houses would be at least twenty-five feet from the dune. The project's initial cost was estimated at $32 million divided among the federal government (70 percent), the state (21 percent), and Suffolk County (9 percent).

As the project neared completion, its backers praised it as proof that with enough will and enough money, people can "hold" the coast against the ocean. For others, the suit established that the government must pay for damage it causes. "This is a landmark case because it puts government on notice that when mistakes happen, you can't turn your back on the families that you damage," Gary A. Vegliante, the mayor of West Hampton Dunes, said at the time.[3] By 1997 the agreement had given him and his fellow homeowners a new beach and the thirty years of relative security as sand is replaced, at government expense, as it erodes.

But for many environmentalists and coastal scientists, West Hampton Dunes is a dismal example of what happens when people disregard the realities of the coast, build on an unstable landform, and then try to avoid the inevitable consequences by armoring the beach. To suit their own purposes, they jettied inlets rather than letting them close naturally to restore the natural flow of sand. When erosion worsened, as it inevitably would, they built groins. When the groins made the problem even worse, they turned to federal, state, and local governments—that is, to the taxpayers. The taxpayers, in turn, find themselves committed to spending tens of millions of dollars to protect the property of people who, in many cases, had every reason to know that they were building or buying in an unsafe place and who, in most cases, were already eligible for additional assistance from the federal flood insurance program. According to initial estimates, the eventual cost of the project works out to about $300,000 per house in the damaged area—about what it might have cost then to buy them out.

Once the project was approved, though, property values began to rise. "The market is beginning to heat up," said John J. O'Connell, a lawyer for the property owners.[4]

What will happen when the thirty-year project at West Hampton Dunes comes to an end? "In theory, in year 31, long after the time when most here will have any concern about the area, that project will terminate," O'Connell told a conference about the Long Island shore in 1994. "But we're betting for our heirs that having spent $80 million and having created one of the most beautiful public beaches in the United States that, thirty-one years from now, government is not going to walk away from that. For one thing, our heirs will be there to see that the public beach continues to be maintained."[5]

Shortly after the suit was settled, almost two years after West Hampton Dunes's near-death experience, the Governor's Coastal Erosion Task Force issued a report acknowledging that people's efforts to control the South Shore of Long Island had caused much of the erosion there. It proposed a number of steps to undo the damage: pumping sand around the inlets, pumping sand onto beaches up and down the coast, and other measures. Ironically, it also proposed that sand be pumped around the Westhampton groin field. That is, it proposed that more equipment and money be enlisted to undo what the groins are doing only too well.

That suggestion raised a question: why not remove the groins altogether? The answer is complicated. First, it is far more difficult, and therefore more expensive, to remove a groin than to construct one. Second, the property owners behind the miles of beach protected by the groins would set up a howl, probably in court. They are the only ones directly benefiting from the groins, but their benefit is substantial—a substantial beach.

Though the story of West Hampton Dunes is extreme, it is far from unique. Over the years, engineers and developers have attempted to trap sand on beaches using everything from stone breakwaters to artificial seaweed. They have hardened the coast with everything from rocks to blocks of petrified sewage sludge. But the interaction between coastal engineering projects and sand on the beach is highly complex and, as the engineers say, "site-dependent." The cost of misjudgment is high. Almost always, these stabilizing and hardening projects have carried high costs and awful consequences.

Though the malign effects of sand-trapping efforts differ, they share the same weakness: they do not add sand to an eroding coast. Instead, they usually work by stealing sand destined for someone else's beach.

Groin construction is the most common—and possibly the least understood—method of sand-trapping or stabilization. Groins, especially the ones built ad hoc, without benefit of engineering, can be quick, relatively cheap, and effective solutions, at least for the person who installs them. They may be built of timber, steel, concrete, rock, or other materials. But seemingly minor changes in their height, length, spacing, and permeability can greatly alter their effects. If they don't trap enough sand, they don't work. If they trap too much sand, they overflow and dump the excess sand into deep water, where it may be lost to the beach.

Usually, groins cause more problems than they cure. They aren't much to look at and they make long walks along an unobstructed beach a thing of the past. On beaches with any kind of surf, they pose a potentially serious hazard to swimmers. And they leave property owners downdrift bereft of sand. Usually, these property owners respond in the only way they can, by constructing their own groins and sending the problem down the coast to the next guy. As one bitter seaside joke has it, groins are like lawyers: once one person has one, everyone else needs one.

This insight is hardly new. Groins—or groynes—were recognized as a problem almost a century ago. "Groynes of all types and sizes and more than thirty in number had been in existence along the Coney Island shore for many years," an engineer wrote in a professional journal in 1923. Yet, he went on,

> taken as a whole, the beach had been eroded, losing at some points substantial amounts. The groynes did in some cases perform the service of preventing further erosion, but that is all that can be claimed for them. As a rule each groyne that was put in place by an upland owner made a small amount of sand in one interior corner or the other but always at the expense of the adjacent owners' land by shutting off the natural sand supply. The best protection that an extended natural beach on the shore can have is the sand itself. Al-

most every beach taken as a whole appears to be injured by jetties or groynes.[6]

The truth pointed out in this article—that the best protection a beach can have is its own unfettered self—is today universally acknowledged. But it is no match for the financial incentives for development on the coast. So when groins began to fall out of favor, property owners and engineers turned to a variety of other devices to trap sand, either in the water or on the beach.

Devices that sit in the water take advantage of a basic principle of hydrodynamics: the faster water is moving, the more sediment it can pick up and carry. When water moves quickly over sediment—such as sand—it picks up particles and carries them along, until it reaches its "carrying capacity." When the water slows, its carrying capacity falls and the particles drop out. Breakwaters operate on this principle. Though they may be underwater, especially at high tide, they are built high enough to break the waves rolling in. The breaking slows the water's motion, causing it to lose carrying capacity. It cannot pick up any more sand, and sand it is already carrying may fall to the ocean floor where, the theory goes, it contributes to the health of the beach.

Often, designers hope that enough sand will escape their breakwaters to ensure that areas downdrift are not severely affected. Often, things do not work out that way. It is not unusual for a breakwater to trap so much sand that the beach builds out to meet it, as has happened in Santa Barbara and Santa Monica, California, and a host of other places. Advocates

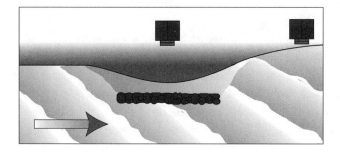

Breakwaters trap sand, in effect, by slowing the movement of water on shore. Sand carried by the water drops to the bottom. *(Jen Christiansen)*

of offshore breakwaters applaud this accumulation of "free" sand, seemingly donated by Mother Nature. The question is, what is happening downdrift, where the sand would have traveled had it not been trapped? "Of course they work," fumes Orrin Pilkey of Duke University, their ardent foe. "They are stealing sand from somebody else."

As more and more construction piles up on the coast, and the financial stakes rise, engineers are devising new varieties of breakwaters they hope will trap sand without causing trouble elsewhere. One company, Breakwaters International, of Flemington, New Jersey, has installed three of its Beachsaver systems on beaches in New Jersey, where hardening and erosion have gone hand in hand longer than anywhere else in the United States. The $2.1 million project was financed by the state and federal governments and by the towns of Avalon, Belmar, and Cape May Point, where the artificial reefs were installed just offshore.

The idea is that incoming waves will expend some of their energy on the reef, rather than on the beach. At high tide, the reefs are designed to be under about six feet of water, so in theory their effects would be most noticeable in storms. The system comprises one hundred ten-foot-long interlocking modules of precast concrete, treated to be salt- and water-resistant. Each module—a triangle in cross section—is six feet high and fifteen feet front-to-back at the base, and weighs twenty tons. The modules, installed by cranes mounted on barges and linked by mortise-and-tenon joints, are positioned offshore. The seaward face of each module slants up in steps designed to return a small amount of sand to the toe of the module, to prevent the intense undermining known as "scouring." The landward side is curved upward and designed to channel backwash into a "flume" that jets up, such that the sand it carries is picked up by incoming waves and carried toward shore by the next wave, especially in storms. The system's designers say that this jet also serves to block movement of sediment seaward during storms.

It may take decades for the effects of such an installation to become apparent on the beach. So far, results have been uncertain or unfortunate.

The first installation was in Avalon in July 1993, on a stretch of beach running from the jetty at the north end of town, at Townsends Inlet. This inlet has been highly unstable (it's not for nothing that Avalon begins at Eighth Street) and skeptics predicted that it would prove impossible to anchor the reefs in the surf near its mouth. Initially, at least, the skeptics were correct. Just as people who stand in the surf will sink ankle-deep as waves swirl sand around their feet, the reefs settled into the sand on which they rested—by as much as five feet at the south end. Perhaps as a result, the reef did not perform as its designers and town officials had hoped. Sand came and went on the beach area protected by the reef, but overall, between the the the installation of the reef and a preliminary evaluation in February 1995, the beach lost sand. Meanwhile, too much sand was accumulating seaward of the reef, an unexpected and unwelcome development. Finally, there was a noticeable problem with scour on its southern end.[7]

The second and third reefs were installed in the summer of 1994. To prevent scour, each was installed on a stone "mattress," topped with a layer of geotextile fabric. The reef at Cape May Point was installed in June, the one in Belmar in August. Both lie between groins on armored beaches. Their long-term effects remain to be seen, but within a year, there was already evidence that the reef in Belmar was sinking, according to Norbert Psuty, a coastal geologist at Rutgers University and an expert on coastal processes.

Engineers designed the Beachsaver Backwash Flume to suspend sand so that currents would deposit it on shore. Divers position one of the components of the New Jersey coast. (*Paul Finkel Associates, Iselin, N.J.*)

In 1988, another kind of reef, designed by American Coastal Engineering of West Palm Beach, Florida, was installed in Palm Beach by Willis du Pont, a local property owner and an heir to the Du Pont chemical fortune. The "prefabricated erosion prevention" or "pep" reef, installed at a cost of $200,000, is triangular in cross section, like the Beachsaver, but taller. Mr. du Pont's beach grew, and the town later received permission to install a larger reef, which resulted in some accretion of sand, including an accumulation at the base of a nearby seawall. But this installation, too, was plagued by settling problems.

Opponents of coastal armor have taken to the op-ed pages of newspapers in New Jersey and elsewhere to criticize the reefs. Their manufacturers and marketers have responded in kind. When Pilkey derided the Breakwaters International reefs as "an old ploy with a high-tech face lift," Professor Michael Bruno of Stevens Institute of Technology in Hoboken, who tested the design in the institute's wave tank for its designers, fired back with a salvo of his own in an article in the *Ocean City Sentinel Ledger*. He said the designs had been described at scientific meetings where "the technologies and the results have received the approval of the coastal community" and that Pilkey's complaints were "grossly inaccurate."

Still, the long-term effects downdrift remain unknown. And although it is too soon to say for sure, submerged artificial reefs may also be associated with safety problems. There have been reports of unusually strong rip currents—and drownings—near artificial reefs, and one report that a man drowned trying to save a child trapped seaward of a partially submerged reef.[8]

The most bizarre form of artificial breakwater is borrowed from nature. The idea grew from the realization that some West Coast beaches bordered by offshore kelp beds suffered less from erosion than other beaches did, apparently because the seaweed slows the movement of the water, causing sand to drop to the bottom. But while natural seaweed appears to accomplish this task easily, artificial seaweed has so far been a bust. When it was tried on the Outer Banks of North Carolina, the plastic fronds broke loose from their moorings, became tangled in fishing nets and propellers, and washed up on the beach. The experiment was deemed a failure.

All kinds of things have been installed up on the dry beach in hopes of trapping sand. One of them, also a product of New Jersey coastal ingenuity, is called the Stabler Disc Erosion Prevention System, sold by the Erosion Control Corporation of Livingston. The system comprises a chain of three concrete discs, each four feet wide and eight inches thick, with steel rings, or slip collars, on either end, mounted on wooden pilings driven into the sand. These chains are installed across a beach above the high tide line where, in theory, they will blunt storm waves and trap sand. Once the trapped sand starts to bury them, laborers must hoist them a bit higher on the pilings so they can continue their work as the beach rises beneath them. If storms erode the sand around them, they simply slide down the pilings and start over. The system was installed on a stretch of more than one thousand feet of beach, the area between two groins, in Spring Lake, New Jersey in 1993. Nearby areas set off by groins were used as controls. Researchers who monitored the results for the company found initially that there was more accretion in the protected groin cell than in the one directly to the north, but accretion to the south was about the same. A year later, they found the protected area had gained about four times as much sand as areas on either side.

If the discs were trapping sand that would otherwise, somehow, have traveled straight offshore, they were working with minimal damage to neighboring beaches. If not, though, the disc system has the same harmthy-neighbor risk as any other sand-trapping system. Also, the discs take up space on the beach and, while not exactly unsightly, they are not the kind of thing people want to see at the beach. When the discs trap sand and start to disappear, they can represent a hazard to unwary strollers. But as Gary Pieringer, the company's vice president for sales and marketing, told a group of scientists led by Psuty on a field trip to the installation, Jersey folk are used to foreign objects on their beaches. "This beach already has groins and pilings," Pieringer said. "People are aware there are hidden hazards here."

In an ideal world, he said, the devices might be installed only in winter, when beachgoers are fewest and the threat of beach loss is greatest. But the system cost about two hundred dollars per foot to install, he said, and towns would probably consider the expense of putting them in and pulling them out every year prohibitive. In fact, he said, the process of raising the discs on the pilings as they become covered with sand is so laborintensive "that they don't always raise them as often as they should."

Still other devices, conical concrete rings invented at Tulane University, were installed on Shell Island, on the Gulf Coast of Louisiana near the mouth of the Mississippi River. The "Beach Cones," ninety-two-pound concrete doughnuts about three feet wide, survived Hurricane Andrew's direct hit at the Louisiana coast in 1992, but the extent to which they helped preserve the beach is not fully known. The devices were also installed at Fourchon Beach, not far away, after the hurricane. Sure enough, the beach widened. But without more study it was impossible to say for sure that the devices had caused, or even contributed to, the accretion.

Engineers seeking still other ways to trap sand are studying the relationship between beach groundwater levels and sand movement, looking to capitalize on the fact that beaches where the water table is high generally erode faster than dryer beaches. The phenomenon was discovered in the 1940s and has been studied sporadically ever since. The idea that a wet beach would erode relatively faster than a dry one makes sense, because it would theoretically be easier to get the sand into suspension—and on the move—if it is already wet.

Companies have sprung up in the United States and in other countries to make and market beach "dewatering" systems, most of which involve filters, pipes, and pumps buried in the sand. Below-surface water passes through the filters into the pipes and is pumped to collection points where it is discharged into the ocean or filtered again for use elsewhere.

The danger, of course, is that this method of keeping sand on a beach may have the same nasty downdrift effects as trapping it by other means. Robert G. Dean, a professor at the University of Florida and one of the nation's leading coastal engineers, studied a Stabeach dewatering system installed in 1988 at Sailfish Point, just north of St. Lucie Inlet, on the east coast of Florida. After two years, Dean said, the segment of beach containing the system appeared to have accreted moderately, while nearby reaches either eroded or accreted slightly. "It is difficult to separate conclusively natural beach changes and those changes caused by the system," he wrote in a report evaluating the system.[9] Similar systems have been installed elsewhere, but have yet to be thoroughly studied long-term.

People such as Pilkey often refer to marketers of shore protection systems as "snake oil salesmen" profiting from the desperate desire of coast-dwellers to preserve their property against the encroaching sea. It is far from clear that such a harsh appraisal is warranted. On the other hand, it is far from clear that anything sold so far actually does what it claims without damaging neighboring beaches, especially over the long term. In a recent statement on research needs, the Marine Board, a branch of the National Research Council, said these devices are "in a class of structures of an experimental nature whose performance cannot at this time be predicted to a reasonable degree. . . . They typically have been sold directly to individual property owners or to communities with erosion problems who were not equipped to make engineering evaluations of the potential or of the actual performance of these products, for the intended application." These communities, the board went on, "find arguments that a device offers permanent protection very attractive, when compared to the continuing investment in maintenance that is required by replenishing the sand supply. When the problem is continuing loss of sand—and this is almost always the case—there is no permanent structural solution that will result in maintaining a recreational beach."[10]

All of these systems have the same crucial failing: all they can do is trap sand. There are places where that is not so bad—like parts of the California coast where sand would otherwise fall into the deep submarine canyons that approach close to shore, or at places like the very western end of Long Island, where sand would otherwise tumble into the underwater canyon carved eons ago by the Hudson River. Once in these canyons, the sand is lost to the beach, so trapping with a so-called "terminal groin" does no harm to anyone. In other places, such as Fisher Island, a private enclave in Miami's harbor, the entire environment is artificial. The sand on this island's beaches arrived by truck in the first place, and keeping it there with groins does not harm beaches on nearby islands.

But these places are relatively few.

Far more common are beaches like those in New Jersey. According to Psuty, hardly any fresh sediment enters the state's coastal system anywhere. The few rivers that might provide some sediment end in sand-trapping estuaries. For New Jersey, beach sand comes from erosion or sediment already in the system, either on the beach or moving in shoreline currents. Trapping sand in these currents only keeps it from reaching beaches downstream. A successful groin, breakwater, or reef system inevitably exacts a price downdrift.

"We are working with a cliff coast," Psuty explained to the scientists who toured the Beachsaver installations with him. "This is the upland," he said, gesturing to the street that ran along the beach, and the homes behind it. "There is no other source of sediment."

In 1995, an expert panel convened by the National Research Council, the research arm of the National Academy of Sciences, reviewed this kind of "nontraditional" shoreline engineering in connection with its study of beach renourishment. Panel members were reluctant to do anything to stifle technical innovation. But they urged that novel devices be tried only after extensive testing, and not simply with models constructed in wave tanks but also in the field. Their report also said that the testing period must be long enough to ensure that long-term performance is not masked by short-term variation.

"No device, conventional or unconventional, can create sand in the surf zone," the committee reported. "Any accumulations must necessarily be at the expense of an adjacent section of the shore." These devices may be beneficial, the report went on, but if they are not, "any unfavorable conditions that develop could be difficult and expensive to correct, including the necessity of removing devices that do not perform well or become hazards to beach users. Further, in the committee's view, some non-traditional devices have been oversold and, with respect to their performance, have shown no lasting capability for shore protection."[11]

The other approach to coastal armor is hardening—walling off the sea. People have been trying to hold off the waves for centuries, and their efforts have resulted in structures of all shapes and sizes, ranging from engineering feats like the seawall in Galveston to rows of sandbags protecting a modest beachfront motel. When a sudden storm threatened to topple the Cape Hatteras Lighthouse in North Carolina, where the beach had been weakened by erosion from an updrift groin field, quick-thinking workers at the National Seashore saved it with slabs of asphalt from a parking lot project, which they threw into the surf crashing around the lighthouse base.

In general, seawall design depends on how much wave energy the wall must hold off. The lightest devices are revetments, designed to protect an embankment from light wave action or currents. These may be simple layers of poured concrete lining the slope, interlocking concrete blocks, or the quarried stone rip-rap. Unless it is constructed on wide, flat, hard shelves, a rip-rap wall will settle and sink in even a small storm, reducing the wall's effectiveness and possibly causing it to fail altogether. Bulkheads are usually larger and more durable; their function is to protect the fill behind them from erosion by larger waves. They are usually vertical walls anchored by pilings, or they may be metal "sheets" of piling, twenty feet wide or wider, driven into the sand side by side. This wall of sheet piling may be reinforced, in turn, with a layer of rock rip-rap laid at its base to prevent scouring. All seawalls are subject to collapse under assault from the waves, but bulkheads are particularly vulnerable. If too much water washes up behind them, they collapse forward onto the beach.

The largest structures are seawalls, massive structures built to resist the full onslaught of the ocean. Some, like the giant wall along the northern

New Jersey coast in Monmouth Beach and Sea Bright, are "rubble mounds"—piles of boulders placed directly on the beach. Or they may be elaborate structures like the Galveston wall, with footings driven deep into the earth and stepped or curved faces designed with elaborate care to reflect the energy of the waves back into the sea. These are usually backed by bulkheads and footed with rip-rap, as is the wall at Galveston. Still, even they are vulnerable to undermining.

Seawalls damage virtually every beach they are built on. If they are built on eroding beaches—and they are rarely built anywhere else[12]—they eventually destroy them. According to Pilkey, "The ability of beaches to retreat landward and build seaward in response to changes in sea level, storm waves, and other natural processes is fundamental to their protective role as well as to their continued existence. Shoreline hardening to thwart nature's ebb and flow is therefore the antithesis of beach conservation."[13]

If the beach is eroding, a seawall will cause it to disappear. A wall by its nature draws a line in the sand. But the ocean does not respect this line. It keeps moving in. Eventually it meets the wall. Result: no beach. As Leatherman put it, "If you have a stable line and you have a moving line, sooner or later they are going to converge." This is the scenario that played out in Galveston and it is happening in almost every other walled beach in the United States, especially on the East and Gulf coasts, where sea level is rising fastest.

Walls built at the base of cliffs last longer and may prevent the cliffs from eroding, but, as with the cliffs at Montauk, that erosion is what feeds the beach when storms or everyday currents rob it of sand. Where there are miles of cliff-backed beach and only a few walls, their effects may not be noticeable. But, as is the case in many parts of the West Coast, small stretches of beach are more or less isolated between rocky headlands. Walling off the sand supply in these cliffs can have disastrous consequences for the beach.

Through a process imperfectly understood, erosion is worse at the ends of seawalls, where they often experience severe scour. In part, this phenomenon may occur because the wall cuts the beach off from a source of sand. The underwater profile at the wall must steepen if the amount of sediment in the water is to remain constant. As a result, many walled towns have repeatedly had to extend their walls when these "end effects" began to threaten neighboring property or the survival of the walls themselves.

There is a lively debate among coastal engineers and geologists over whether seawalls themselves accelerate the erosion overall on the coasts

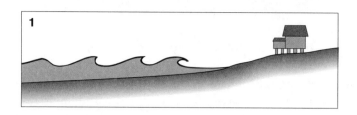

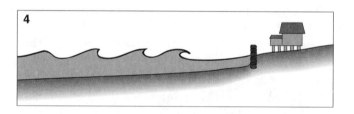

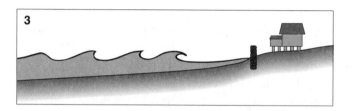

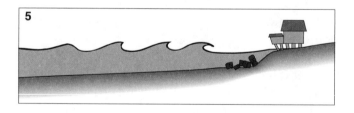

If a shoreline is eroding (1), building a seawall will destroy the beach. First, the wall itself reduces the size of the beach (2). As water moves in (3), it eventually meets the wall, flooding the beach (4). Ultimately, wave action undermines the wall, causing its collapse (5). *(Jen Christiansen)*

they front. In 1990, a report by the National Research Council concluded that "properly engineered seawalls and revetments can protect the land behind them without causing adverse effects to the fronting beaches."[14] Even allowing for the fudge factor inherent in the phrase "properly engineered" (and some critics argue that if bridges failed as often as seawalls no one would cross water except by boat), the conclusion has provoked bitter argument from coastal geologists opposed to hardening.

Their theory goes like this: when a wave runs up a natural beach, its energy is gradually vitiated as water soaks into the sand. By the time the wave finishes its run-up, it has little energy left to carry sand away when it runs back out to sea. By contrast, the theory goes, waves that strike a seawall bounce right back with much of their energy intact, moving fast, stirring up the bottom, picking up grains of sand and carrying them off. As a re-

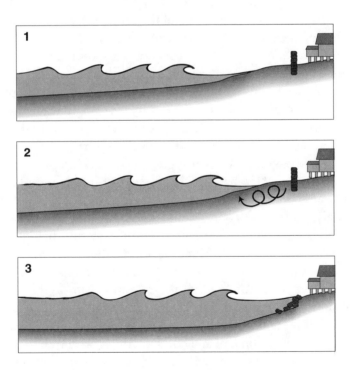

Some scientists believe that seawalls encourage erosion. They theorize that when a wall is built, storm waves are no longer absorbed by the sand (1). Instead, they bounce off the wall, scouring the beach in front of it (2). The wall eventually fails (3). *(Jen Christiansen)*

sult, the base of the wall is scoured and the wall itself is threatened. Though this theory has yet to be proved, this kind of scour has haunted seawalls of all descriptions.

New Jersey is home to the nation's oldest beach development and, as a result, has paved the way for many of the worst abominations along the nation's coast. In effect, the people of New Jersey have been conducting a giant field experiment in shoreline stabilization for more than 125 years. Today, the state has one of the most heavily armored coastlines in the United States; in this century, more than three hundred groins and scores of walls, revetments, bulkheads, and other structures have been built to keep the beach in place.[15] The result is a coast in crisis.

The first seawalls were built early in the twentieth century to stabilize the eroding shore and protect the road along the coast and the buildings fronting it from storm damage. One of the longest and tallest stretches of seawall in New Jersey—indeed, anywhere—runs along Sea Bright and Monmouth Beach, two towns on a barrier beach south of New York City. Where once there was a beach hundreds of yards wide, storm waves crash on a wall of rocks. Even coastal engineers call the wall "infamous."[16]

In the early part of the nineteenth century, Sea Bright was a narrow, undeveloped sand spit, where a few dozen fishermen set up primitive shacks. The first permanent house was built in 1869, and by 1877 the beach was lined with houses—and its dune was gone.[17] An 1886 photograph shows at least one house protected by a bulkhead; the first rubble mound wall was installed twelve years later. By 1931, the wall had just about achieved its present twelve-mile length and seventeen-foot height.

For much of its length, a four-lane state highway runs behind the wall. Driving in the shadow of this grim, gray fortification is the antithesis of a day at the beach. Along much of its length, the wall towers so far above the road that drivers have no clue that there is an ocean on the other side. Some oceanfront residents of Sea Bright can view the Atlantic only from the second or third floors of their homes; visitors who want to see the ocean must cross the highway on foot and climb up the rocks. From the top of the wall, they view waves crashing on rock. There is almost no beach.

The land protected by this enormous structure is a strip of sand in places less than two hundred yards wide. At the wall's north end, the land is occupied by a row of unimpressive motels and diners; further south, neighborhoods are more elaborate but, not surprisingly, the loss of their beach has caused them to deteriorate markedly. True, the wall protects access to Sandy Hook, the curving spit that marks the entrance to New York harbor and is one of the nation's most heavily used parks. But had the coast been free to migrate by itself, land access would probably remain today (though it might be less convenient). By halting erosion at Sea Bright, the wall cut off the supply of sand that would otherwise have moved north, toward Sandy Hook.

In any event, by 1994, tens of millions of dollars had been spent over the years to maintain the wall and the property behind it. But the Corps of Engineers had even bigger plans: a project to pump millions of cubic yards of sand to the foot of the wall, to restore the beaches lost since the wall was built, as well as other beaches to the south. A year later, people in a high-rise condo and two formerly beachless private beach clubs could look across the wall and out to sand again. But even before the project was halfway complete, a series of storms had washed much of it away, leaving several unanswered questions behind.

The federal government spends millions to maintain this wall, in Sea Bright, N.J., photographed after the Halloween northeaster of 1991. *(The New York Times)*

How long will the new beach last? No one knows. Apparently because of the wall's presence, the underwater profile of the beach has steepened substantially. Only three hundred yards from shore, the water is thirty feet deep or deeper. According to Psuty, the water only gets this deep at other New Jersey barriers a mile or so offshore. If the sand applied to the beach seeks to stabilize in what engineers call an "equilibrium profile," it will soon disappear under the water.

Will there be money enough — $1 billion by some estimates — to maintain the project over its fifty-year lifetime? Already the corps is talking about spending cuts. And is the development protected by the wall and its new beach worth the cost? For a few, mostly people with business or real estate interests in the area, the answer to that question is loud and affirmative. For the rest, the answer is no.

A similar loss of beach befell Cape May, a Victorian beach resort at the southern tip of New Jersey, after the town built a seawall to combat the effects of jetties the Corps of Engineers had built at Cape May Inlet to the north. In much of Cape May today, a beach stroll is a walk atop a wall of boulders.

In 1989, Professor Mary Jo Hall of Rider College in Trenton, New Jersey, mapped 108 miles of New Jersey coast and marked jetties, groins, seawalls, bulkheads, revetments, submerged breakwaters, and other devices. On average, she found, beaches with no stabilizing devices or structures had 180 feet of beach from the high tide line to the dune line, the area geologists call the dry beach. By contrast, areas downdrift of groins or jetties had an average of only 60 feet; beaches fronting seawalls were, on average, only 29 feet wide. Though there were a few small areas where there was sand only because of armoring, she found twelve miles of armored coast where there was no beach at all at high tide.[18]

Opponents of beachfront armor often cite "New Jerseyization" when they want to describe what happens when buildings take precedence over beaches. Unfortunately, however, people in other states have failed to learn from New Jersey's sad example. By some estimates, almost 50 percent of the nation's shoreline is armored with hard structures and, despite new laws aimed at restricting them, they continue to proliferate, with predictable effects.

One example is Sandbridge, Virginia, a small community of modest houses, most ranch style and split-level, on tiny lots lining two or three streets along a barrier spit about ten miles from the North Carolina border. When it was first established, in the 1960s, Sandbridge had the air of a typical suburban development—with one magnificent difference: a sandy beach three hundred feet wide.

But Sandbridge was eroding, and fast. In geological terms, the village is unlucky. The ocean bottom drops off quickly there; geologists say the contour line marking the two-meter depth hugs the shore, and the ten-meter line is only about twelve hundred meters offshore. Deeper water means bigger waves. Unlike people elsewhere, residents did not turn to the government for help, in part because it would have meant opening any rebuilt beach to the general public. Instead, property owners along the shoreline built their own bulkheads. Many simply drove sheet piling deep into the sand in front of their houses and then ran wooden stairways eight or ten feet down to the beach. The work was begun in 1978, when a few property owners built walls. Others quickly followed suit, often because of the end-effects of a neighbor's wall. By the early 1990s, more than half the four-and-a-half-mile stretch designated as Sandbridge was walled off by vertical sheet piling, much of it steel but some concrete or timber.

Sandbridge, Va., seawalls after the Halloween northeaster of 1991. *(Mort Fryman, Duke University Center for Study of the Developed Shoreline)*

The project's success was short-lived. Erosion undermined some of the sheet-pile walls. Others had inadequate drainage. When storms carried water behind them, the pressure of the water forced the walls to slump forward onto the remaining beach.

But the beach loss at Sandbridge has nothing to do with deficiencies in the design or construction techniques and only a little to do with the nearshore slope of its seafloor. The steep slope may worsen the erosion rate a bit, but it does not cause the underlying situation that makes the walls a mistake. Sandbridge is at a zone of high erosion—about three feet a year on the north end to ten feet a year at the south. The sea is moving in on the land, and the walls keep the beach from moving with it. The people of Sandbridge fixed an unmoving line, and the moving sea steadily encroached on it. "We're foolish now that we are this close to the ocean," George Owens, then president of the Sandbridge Property Owners' Association, said when the situation became critical in the early 1990s. "But we didn't used to be this close."

Today there is almost no beach in Sandbridge, and to reach what little remains residents must climb steep stairways that often end in midair a few feet above the sand. Sand that once surrounded them at the base has been eroded away. The height of the walls and the narrowness of the dry beach area combine to make sunbathing impossible in the afternoon, when the narrow stretch of beach is in the shadow of the rusting sheets of metal piling.

A few hundred miles south is Fripp Island, a gated island community on the South Carolina coast that bills itself as a luxury resort. But when the pamphlet describing its upscale accommodations talks about swimming, it talks about pools. For Fripp Island is almost entirely armored and much of it, at times, is beach-free. The island, only about three miles long with unstable inlets at either end, regularly experiences fast-moving erosion. As a result, it has been armored with groins, a jetty, seawalls, revetments, and bulkheads. Fripp Islanders have attempted to fix the position of the coastline of an eroding island. Result: vanishing beaches.

Some residents suggested pumping in more sand, but the consultant they hired offered little encouragement, telling them it would only wash away. Meanwhile, every big storm leaves a few more property owners with failed bulkheads. As the few lots that are not bulkheaded erode, the ocean flanks the bulkheads on adjoining property, threatening their stability. As a result, people who have walled off their property urge their neighbors to do the same.

Ironically, in the early 1990s Fripp Island won a reprieve of sorts when the state of South Carolina undertook a sand-pumping operation on an island to the north, at Hunting Island State Park. Much of the sand applied to the narrowing beaches there washed away, and much of it moved onto the Fripp beaches. But it will not last forever. Developers and property owners have cast their lot with engineering—to do otherwise would have meant abandoning some prime property. Now the island is just about surrounded by armor, and property owners are wondering what to do. *Living with the South Carolina Shore*, one of a series of books published by Duke University Center for the Study of the Developed Shoreline, which Pilkey runs, has alarming advice for people considering moving to Fripp Island: "Avoid beach-front property. . . . In the case of a hurricane warning, evacuate early."[19]

Florida also has its beachless beach towns. One of the most poignant is Boca Grande on the state's west coast. The town, reachable by a narrow causeway and a tiny bridge, lies at the southern end of Gasparilla Island, one of the few places on the Florida coast that retain the air of the more leisurely—and less commercial—past. It is a small town of white stucco houses whose red tile roofs drip with flowering vines, and its major avenue is a shady arcade of ancient banyans. But despite its seemingly ideal location on the Gulf of Mexico, Boca Grande has hardly any beach. Most of the property owners on the Gulf side have installed rock rip-rap or concrete walls to protect their gracious stucco homes. Lawns run lush and green down to their seawalls but, absent artificial renourishment, there is only a narrow strip of sand and that only at low tide.

The lighthouse at the southern tip of the island, a local landmark, is fronted by a modest expanse of sand, but even this small beach is held in place by rock breakwaters. Perhaps because these breakwaters collect sand at the lighthouse, the town beach nearby is only a few yards wide at high tide. Property owners on either end of this beach have walled off their yards, creating a situation not uncommon when walls and natural beaches meet. The shoreline at the town beach, unobstructed by beachfront armor, has retreated naturally, leaving the walled areas sticking out into the gulf, vulnerable to flanking.

The same kind of situation can be seen in Palm Beach, the far more lavish barrier island town on the other side of the state. When the railroad magnate Henry Morrison Flagler took his railroad there in 1894 and began a building boom in Moorish mansions, wags described Palm Beach as "what God would do if he had the money." Today, parts of Palm Beach are practically beachless. Almost every inch of oceanfront is backed by sea-walls. Though the land is high enough that the oceanside walls do not usually block the view of the sea, and though the walls are usually softened with flowering vines or other plants, there is no hiding the fact that the beach at Palm Beach is almost gone. The rack line—the seaweed, shells, and small pieces of driftwood that mark the high tide's reach—runs about ten feet from the wall, where there is any beach at all. Residents of many of the city's lavish mansions must climb metal ladders down cement walls if they want to walk on the sand.

The geology is different on the West Coast, but the effects can be similar. For example, though Puget Sound in Washington is relatively calm and erosion is not a big problem, much of its shoreline is walled. Some walls have been put in simply to provide a "cosmetic edge" to waterfront property; overall, more than 50 percent of the shoreline in King County, Washington, on the sound's west coast, is armored. But beaches in Puget Sound and elsewhere in the Pacific Northwest are short and segmented, and they are formed primarily of material eroded from adjacent bluffs. The amount of sand on the beaches depends not only on the size of the bluffs and their composition but also on the amount of erosion they undergo. Beaches will survive only if there is enough erosion to feed them the sediment they need. If an area has only one eroding bluff, and it is walled off, the entire place is deprived of sediment. The underwater beach profile gets steeper and waves attack the shore more easily. In a kind of domino effect, other property owners are then forced to install their own protective structures until, eventually, the walls themselves are under-mined and the erosion cycle begins again.

The problem has intensified in recent years as the small summer cabins that once dotted the cliffs have given way to enormous, year-round homes near the bluff edge, prompting calls for more and more elaborate erosion protection.

A seawall does not have to be made of concrete to damage a beach. Though few visitors to North Carolina's Outer Banks realize it, the entire string of barrier islands that stretches its narrow way from the Virginia border past Cape Hatteras to Ocracoke is fronted by a seawall. It looks like a sand dune but it is a wall just the same, built by the Civilian Conservation Corps in a project started during the Depression. The project was begun after two hurricanes had struck the Banks in 1933, and its aim was to improve the Banks's economy by holding off the sea and allowing the construction and preservation of a paved road. The project would employ hundreds of workers to "restore" dunes to the natural conditions its planners believed had existed until storms, timber-cutting, and overgrazing destroyed them.

But no one knew what the Banks had been like when European settlers first arrived hundreds of years before. The earliest detailed descriptions of the Banks date from the late 1800s, and by then settlers had been there for more than one hundred years, along with their goats, sheep, cows, horses, and hogs—and their appetite for wood. In those days, the islands were flat stretches of beach and modest dunes that gave way to grassland and salt marsh. But it was widely assumed that when Native Americans inhabited them, the islands had been wooded, with large dunes, and that woodcutting and overgrazing had somehow destroyed the vegetation and weakened the dunes. The artificial dune would help re-create the ancient landscape.

By the 1950s the project was complete. The workers had installed 557 miles of sand fences, planted 3,254 acres of beach grasses, and set out 2.5 million seedlings, trees, and shrubs.[20] The result was a line of vegetated dunes, fifteen to twenty-five feet high, running for almost a hundred miles. But the project was based on a false premise. Coastal geologists now believe that much of the Outer Banks had probably never been a stable, thickly planted landscape protected by high dunes. South of Nags Head, the barrier islands probably had always been low and marshy and, as scientists have since established, they are slowly retreating toward the mainland, as sea level rises. Their principal mechanism for this migration is overwash, as storm winds, great ocean swells, and high tides carry sand across the island. Until the construction of the artificial dune, overwash routinely carried tons of sand into the islands' sound-side marshes with every storm. But now the new dune was blocking overwash except in the most extreme weather, leaving no source of sediment to feed the back side of the islands, which also began to erode.

Paul Godfrey, a onetime research biologist for the National Park Service and by then a professor of botany at the University of Massachusetts, told this horrible truth in a monograph published by the National Park Service in 1970.[21] In the next few years, he and others elaborated on this work. Citing research in the Everglades and in Californian redwood and chaparral forests, where routine fire control measures eventually resulted in highly destructive fires fueled by accumulations of forest debris or other organic matter, they concluded that "natural disruptive change is often essential to the maintenance of ecosystem structure and function."[22] For the Park Service, they went on, this means "that a particular area must be maintained safely for those who wish to use it for recreational purposes, while at the same time the area must be allowed to remain in a dynamic biophysical state." In other words, the frequent overwash that once made a road impossible was the very process the islands needed to survive. In the bland language of science, Godfrey and his colleagues called this situation a "management contradiction." It would come to haunt the Outer Banks.

Though many other coastal geologists could hardly believe the artificial dune could do such damage, the people who lived on the Banks knew Godfrey's theory was correct. One of them was Wayne Gray, a retired coastguardsman who was born on the Banks before the dune was built, and grew up in the small town of Avon before World War II, when it was an isolated village of a few hundred souls, unconnected to any other settlement except by the beach, which people would traverse by horse cart. In those days, he recalled, winter storms would send water rushing under his house, into the marsh.

By the 1970s, three decades of interference with overwash had greatly altered the shape of the beaches on the Banks compared to beaches on islands further south, where there was no artificial dune. On those islands, beaches are 125 to 200 yards wide. Beaches backed by the dune, by contrast, were 30 yards wide or less, and in many places there was hardly any beach at all. Underlying erosion, "combined with the presence of a permanent dune structure has created a situation in which high wave energy is concentrated in an increasingly restricted run-up area, resulting in a steeper beach profile, increased turbulence, and a tendency for the beach sand to be broken up into finer pieces and washed away."[23]

Scientists eventually accepted the startling theory, and the National Park Service decided it would no longer defend the dune. It would let it erode. The Park Service declared:

Following damaging storms, the dunes [will] not be artificially rebuilt, but in extensive barren areas a revegetation program [will] be initiated. Inlets which opened during storms [will] be permitted to migrate and close naturally. This alternative envisions that at some time in the future it may be impractical to maintain a continuous road through the seashore.[24]

But the road had already led to development north of Cape Hatteras, and the Park Service was not the only voice heard there. The state of North Carolina now has an important interest in the dune, because it protects the road, the only way on and off the islands. "You can repair the dune and have your highways and have a normal life," Mr. Gray observed a few years ago, after yet another storm broke through the dune in several places and flooded the highway. "Or you can let it go. And then I would-n't know what to do. In 1950 people could survive without the road. Now they can't."

But, as seawalls eventually must, the dune is beginning to fail. Every time there is a storm, another portion washes out and, despite the Park Service's declaration, the state protects the road with giant sandbags cam-ouflaged with sand. The state has also relocated some sections of highway and would like to move much of the rest. But where? The islands have narrowed so much that in many stretches there is no longer any good place to put a road.

There are plenty of places where natural erosion rates are so modest and the property behind the wall is so valuable that walls make good econom-ic sense, even if a certain amount of beach is lost.

The O'Shaughnessy seawall in San Francisco is one example. The water there is far too cold for swimming, and the wet-suited surfers do not seem to mind that the beach in front of the wall is narrow to nonexistent. In fact, many people in San Francisco say the concrete wall's stepped de-sign has added to, not detracted from, the ocean front since it was finished in 1922.[25] And it certainly beats a previous method of protecting the city's waterfront: dumping rubble into the threatening high tide.

Then there is the Hurricane Barrier in Providence, Rhode Island. Built after hurricane floods had surged into the city's downtown twice in sixteen years, the barrier combines a seawall and tide gate that can be closed when storms threaten. The huge metal gates—almost unnoticeable among the gas plants and oil tanks at the head of the bay—have been closed only a handful of times since its construction three decades ago, but Providence sleeps better because of it and has suffered no appreciable loss of recre-ation. The seawalls that protect the giant petrochemical complex at Texas City, Texas, also cause few feelings of regret. The infrastructure they pro-tect is immensely valuable.

Sometimes the environment in question is so completely shaped by human hands that a little more interference is of no importance. Santa Monica Bay in California is so controlled by groins, breakwaters, and sea-walls that questions of beach aesthetics no longer arise. And because the sand in question would otherwise vanish in a submarine canyon, no one loses when it is trapped in the bay instead. When much of Santa Monica

Bay was armored, beginning in the 1930s, people did not realize what they were doing. According to Inman, the theorist of the littoral cell, "Santa Monica Bay is the principal example of a cell that is now stabilized by actions of man—not intentionally but over a period of thirty to forty years simply by building structures along the coast. Santa Monica Bay in the first place was highly industrialized and the population density is very acute so these things happened. Money was raised to do these things, rightly or wrongly."

Other areas, even nearby, would not do so well, Inman said. As an example he cited Oceanside, a coastal town between Los Angeles and San Diego. "Suppose we did the same thing. The cost would be astronomical." The Oceanside littoral cell is three times as long as Santa Monica's, sixty miles. It lacks Santa Monica's protective headlands, leaving it wide open to waves from all directions. The only stabilization is at Oceanside Harbor in the middle of the cell, where jetties have caused problems for years by interrupting the flow of sand.

"I am neither advocating beach stabilization or strongly supporting natural conditions," Inman said. "I am simply saying you have to look at the overall problems and make rational policy decisions. Because it's not possible to say 'I am going to stabilize this middle section and not stabilize anything else.'"

Erosion does not threaten the beach per se. Left to confront a rising sea alone, a beach will simply move inland. Decisions to armor the coast are not decisions to save the beach—quite the contrary. They are decisions to sacrifice the beach, or a neighboring beach, for the sake of buildings. Unfortunately, this thought usually gets lost in the panic when quick action is urged to prevent a house or motel from falling into the surf. The case for shoreline hardening "must stand or fall on the comparison between the costs of a structure and the value that it is intended to preserve or produce," the Rhode Island Development Council said as long ago as 1954, in a report written after yet another hurricane brought a chorus of calls for coastal armor. "Adequate protective structures, especially seawalls, are expensive and can only be justified where property values are high or some other impelling reason exists."[26]

But calculating these costs is not easy. On the plus side, in theory, is the value of the buildings to be protected. In some areas, this value comprises tourism jobs and possibly a substantial portion of a region's economy. On the other hand, this value usually derives in large part from the buildings' location—that is, from their proximity to the beach. If shoreline armor costs a community its beach, it is a net drain, not a net benefit. And that is only one of the weights that must be added to the negative side.

- Armor degrades beaches, most obviously through impoundment or passive erosion. "If you cover a beach with rock, it's gone," says Gary Griggs, a geologist at the University of California at Santa Cruz who has studied the effects of seawalls. Perhaps because they are concentrating on buildings, not sand, engineers do not usually take passive erosion into account when evaluating the performance of a seawall.
- Armored beaches are unsightly, especially compared to natural beaches.
- Walls, groins, and breakwaters are dangerous to people who use the beach. Swimmers must keep out of their way and beach strollers may find themselves rock-climbing instead. It is not unusual for rocks in a rubble mound wall to roll onto people's fingers, for climbers to twist or even break ankles on the rocks, or for rising water to trap unwary sunbathers against rock walls. Underwater breakwaters are hazards to navigation.
- Hard structures reduce access to the beach, both for people who want to use it and for those whose view of sand and surf is replaced by one of rock or concrete.
- Once stabilization starts, it tends to spread as one groin, breakwater, or wall accelerates erosion downdrift.
- Armor is expensive to install and expensive to maintain. Coastal armor becomes a part of a community's infrastructure, like its roads or water system, and like them it must be maintained in perpetuity. Many communities are finding that the best way to protect their armor is to build an artificial beach in front of it by dredging sand from offshore or a nearby inlet and pumping it onto the shore. But these beach nourishment projects are themselves expensive and must be repeated often.
- Armor benefits a comparatively small group of people as compared with those who might use the beach. (Often, the beneficiaries are not the ones who pay for it.)
- When beaches are lost to armor, birds, shellfish, and other creatures lose their habitats.
- Armor creates a false sense of security that can encourage even more irresponsible construction.

Many coastal scientists hope that the evil effects of shoreline hardening are so well known by now that the strategy will be abandoned as a technique for saving endangered coastal buildings. Some states—including

North Carolina, Maine, New Jersey, and South Carolina—have already taken steps to prohibit further hard stabilization; to one degree or another, they bar construction of revetments, bulkheads, seawalls, groins, jetties, or other kinds of armor. Other states, like Florida, have restricted their use. "We're probably not going to build too many more seawalls," Leatherman told the Woods Hole retreat sponsored by the National Academy. "People look at them as a last resort."

But even as he was offering this optimistic assessment, property owners up and down the coast were challenging it, appealing for permission to build just one small stretch of wall or one small groin to protect their home or business. In a contest between the abstract idea that seawalls damage beaches and the all-too-certain reality that one's house is about to fall to the sea, reality wins out. People whose structures are at stake want to save them. They want to take refuge behind seawalls, bulkheads, rock revetments, and the like. Faced with the choice of saving building or beach, few are brave enough—or farsighted, benevolent, or rich enough—to choose beach. They want walls now. They will worry about the beach later.

Those who love the Montauk Lighthouse devised a plan to save it. Unfortunately, their plan was to armor the base of the cliffs with boulder riprap. The plan took shape in 1991, after the Halloween storm ate thirty feet of cliff away. The first shipment of ten-ton boulders arrived in 1992, financed by donations from East End celebrities, a benefit concert organized by the singer Paul Simon, and ordinary people who paid three hundred dollars each to put their own rock at the base of the cliff.

When a devastating northeaster struck in December 1992, the *East Hampton Star* reported the damage in words and pictures, including photos of intense beach erosion. Tucked unobtrusively into a small box in the bottom of another page was a happier item, or so it seemed. The erosion control project at the Montauk Lighthouse "got its first test last weekend and appeared to pass with flying colors," it said. The project—rock rip-rap along two hundred feet of bluff at the south-facing side—had taken a direct hit, but, as Greg Donohue, designer of the project, proclaimed, "there was no shifting of stones. We were tested and we passed, thanks to Paul Simon and all the people who stood in line" for his concerts.

But no sand would be moving west to feed Long Island's starving beaches.

CHAPTER FOUR

Unkind Cuts

A small rock holds back a great wave.
—Homer, *The Iliad*

On an isolated stretch of Oregon coast, just south of Tillamook Bay, a dirt track runs off the paved road into the scrubby vegetation of a narrow barrier spit. A sign stands at the corner. This is what it says:

> City of Bay Ocean Park
> In 1906 T. B. Potter, real estate broker from Kansas City, dreamed of this peninsula being a second Atlantic City. Francis B. Mitchell bought the first lot in 1907 and he was the last to leave in 1952. Business had a grand opening June 22, 1912. It consisted of a general store, post office, a three story hotel, bowling alley, tin shop and bakery. The hotel had automatic fire sprinklers. There was a natatorium with a pool 50 x 160 feet. There were 4 miles of pavement, city lights and water, telephone system and a narrow gauge railroad. By 1914, 600 building lots had been sold and 2,000 people involved. Due to erosion cutting banks the natatorium was destroyed in 1932. Over 20 houses had fallen into the sea by 1949. In 1952 the sea cut a half mile swath making Bay Ocean an island. Breakwater was built in 1956 and reestablished the peninsula. Of 59 homes and summer cottages only 5 were moved in time. On Feb. 15, 1960 the last house was washed into the sea and the city of Bay Ocean was a dream again.

This sign, placed by a local chapter of the American Association of Retired Persons, is all that remains of Bay Ocean Park, a town destroyed by a seemingly beneficent bit of coastal engineering.

The story begins with Thomas Irving Potter, the son of the Kansas City developer Thomas Benton Potter, who took a vacation on the northern coast of Oregon in 1906. One day, young Potter and a friend went boating in Tillamook Bay, west of Portland. As they neared a peninsula at the mouth of the bay, Potter spied and shot a goose. It fell to earth on the uninhabited spit, and the pair landed their boat so Potter could climb up a dune to retrieve his prey. When he reached the top, a glorious vista spread out before him: a narrow peninsula four miles long, with sand dunes more than one hundred feet tall, forested with spruce, cedar, and juniper. A wide beach ran its entire length. When he returned home, Potter told his father about the wild landscape. T. B. Potter resolved at once to buy the site and build a resort he would promote as "the Atlantic City of the West."

By the end of the decade, the elder Potter's development was well under way. Because the peninsula lay between Tillamook Bay and the Pacific Ocean, he called the place Bayocean. When the resort had its formal grand opening on June 22, 1912, it boasted hotels, cafes, a bakery, an outdoor pool, and an indoor "natatorium" pool complete with electrically powered surf. By 1914, there was railroad and ferry access and sixteen hundred lots had been sold, many to people who lived in the town year-round. Early postcards show young women in broad-brimmed straw hats and long white muslin dresses strolling on the wide beach.

By then, however, plans had already been made that would doom the community. Businesses in towns on the north side of Tillamook Bay, across a narrow inlet from Bayocean, were eager to improve boat access and upgrade their harbor to handle something more than their major cargo, low-grade logs. But the inlet was clogged by shoals of shifting gray sand.

Town officials appealed to the Army Corps of Engineers, and the corps responded with a plan for two jetties, one on each side of the inlet. The jetties, long rock fingers lining the edges of the inlet and sticking out into the ocean, would trap sand moving north or south and keep it out of the harbor mouth. The project would cost $2.2 million, of which the towns would have to pay 25 percent. Daunted by the cost, the communities instead suggested that only one jetty be built, on the north side of the inlet, and that some dredging be done to keep the inlet clear. This work, they estimated, would cost $814,000.[1] Though corps officials said they could not predict the effects of building only one jetty—usually they were built in pairs, one on each side of the inlet mouth—Potter was enthusiastic. The project would make the harbor one of the safest deepwater anchorages between San Francisco and the Columbia River, he wrote in his newspaper, *The Surf*. The region would be "a mecca for tourists."[2]

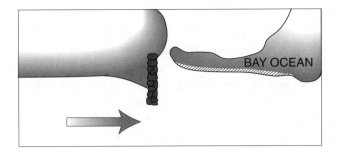

Until the Corps of Engineers built a jetty at Bayocean, Ore., currents carried sand up and down the coast. The jetty trapped sand coming from the north, causing the spit at Bayocean to erode. *(Jen Christiansen)*

The north jetty was completed in 1917. Almost immediately, sand began to pile up behind it. As quickly, residents of Bayocean began to see their beach disappear. By 1920, oceanfront property owners were moving their houses back from the encroaching sea. The jetty was repaired and extended in 1932 and 1933. Erosion at Bayocean intensified. By the mid-1930s, waves were eating away at the town sidewalks and the foundation of the natatorium, which had been built right on the dune line and was too big to move. Its roof caved in 1936, and a storm three years later wrecked the rest of it, leaving only chunks of concrete on the beach.

In 1948, another severe storm struck, breaching Bayocean spit and severing water and electric lines. More than twenty houses were lost to the sea, along with hundreds of feet of beach. Some people who had not built on their lots simply abandoned them; Tillamook County claimed them for taxes, but many were already underwater. The spit breached again in 1952. The Bayocean Post Office, opened with some fanfare in 1909, closed for good in 1953. The hotel was abandoned to squatters; eventually, it and all the other structures on the island yielded to the sea. The last house slid into the ocean in 1960.

Today, the dirt road runs to the end of a greatly reduced spit of land. The dunes and tall trees are gone. In their place are small spruces, Scotch broom, Queen Anne's lace, and grass, mute testimony to what can happen when people interfere with nature on the coast.

People have tried to control inlets for centuries. In earlier days, they dug by hand to keep inlet channels deep enough for boats. Usually, these efforts failed. For example, men with shovels tried in 1830 to stabilize Ocracoke Inlet, on the Outer Banks of North Carolina, but the relentless flow of sand-laden currents eventually forced them to accept the fact that the inlet would always be shifty and full of shoals. In the nineteenth century, residents of Martha's Vineyard attempted repeatedly—and in vain—to open an inlet on the south shore of the island, only to see a storm accomplish the same thing in an afternoon. "The question of controlling these elements by artificial means is one of much uncertainty," Henry L. Whiting, a surveyor with the U.S. Coast and Geodetic Survey, wrote to the superintendent in 1889, in a report about the inlet.[3]

Nevertheless, as coastal engineering improved and the commercial benefits of safe, convenient outlets to the sea became apparent, more and more coastal towns stabilized their inlets with jetties. Today, jetties are among the most widespread features on the coast. They are also among the most destructive. Their very name tells why: they disrupt the currents that carry sediment alongshore and either trap or "jet" it out into deep water. In doing so, they destroy the natural movement of sand along the coast, depriving beaches downdrift of the sand they need to replace what they are losing to littoral drift.

In nature, inlets also interrupt this drift, but in nature, inlets usually heal, migrating and reforming in complex patterns. Powerful hurricanes whose storm surges send vast volumes of water over a barrier can cut deep paths back to the sea in a matter of hours—as hurricanes did when they formed inlets in 1933 in Ocean City, Maryland, and in 1938 at Shinnecock Bay on the south shore of Long Island. In northeasters, the fierce Atlantic storms that occur most often in winter, the process may take longer, as each tide cycle sends water across the barrier and then out again through the breach, widening it. Strong northeasters that linger for days, as they often do on the East Coast, can easily create multiple inlets.

Once an inlet forms, it may linger for years, even decades. More often, however, inlets begin to heal as soon as the storm ends. How fast they heal depends on two factors: the amount of sand in the littoral drift and the amount of water flowing through the inlet with the tide. If currents carry abundant sand and tidal flow is weak, sediment starts to fill the inlet at once and the inlet heals. In places like the Outer Banks of North Carolina, where the tidal range is less than two feet and littoral drift is rich with sand, inlets tend to heal quickly. At various times since the Banks

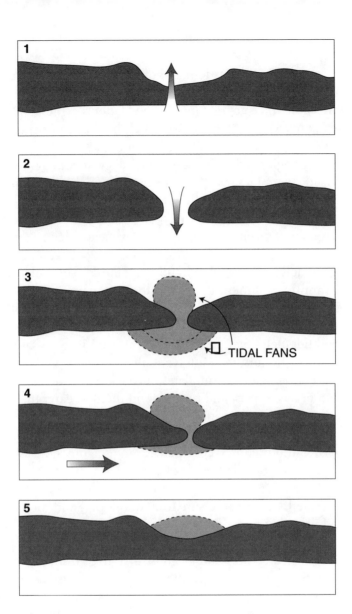

Left to themselves, inlets come and go naturally. Usually, an inlet forms when a storm surge carries water across a barrier island (1). When the storm ends, the excess water seeks a way back to the sea. It finds a weak spot and breaks through (2). Currents carry sand along the island shore, and tides carry it in and out of the new inlet's mouth, where it is deposited in tidal fans (3). The inlet mouth fills with shoals (4), and eventually the inlet heals over (5). *(Jen Christiansen)*

were colonized by Europeans four hundred years ago, thirty inlets have come and gone and in some cases come back again. Today there are three more or less stable inlets there: Oregon and Hatteras, which both opened in a hurricane in 1846, and Ocracoke, which has been present since Europeans settled there in the seventeenth century.

Inlet healing begins as ebb and flood tides deposit sand in lobes just inside and outside the inlet's mouth. Some of the sand deposited in ebb tide deltas as the tide runs out of an inlet may be picked up again by littoral drift, to resume its journey along the coast. (This was the transport mechanism that apparently kept the beaches at Bayocean adequately supplied with sand.)

A flood tide delta, on the other hand, is a natural "sink" of sand inside the inlet's mouth. Sediment deposited there remains relatively undisturbed. Marine plants may colonize it, slowing the movement of the water and causing even more sand to settle onto it. Once a grain of sand is tied up in a flood tide delta behind the barrier, it is unlikely to make its way back out to sea without human intervention.

As more and more sand accumulates, shoals form inside the inlet itself, and its channel has to twist and turn around them to carry the tidewater back to the sea, causing even more sand to drop out. Eventually the inlet chokes on the sand and closes. Inlet formation removes sediment from the beach. But the sand is conserved. In fact, some scientists believe 20 to 25 percent of barrier island volume may be sand left behind in the deltas or throats of ancient inlets.[4]

In the twentieth century, pressure from developers and sport and commercial and fishing interests have interfered with this process. In many places—Ocean City and Shinnecock Inlet among them—the inlets were regarded as gifts from the weather gods, the silver lining in the storm clouds. When Shinnecock Inlet cut through the barrier island on Long Island's south shore, for example, people on Long Island itself suddenly found their boats could reach the ocean more quickly. In areas that have—or want—commercial fishing fleets, the sudden appearance of a new inlet can be a real boon. Also, if waste from inland agriculture or overdevelopment is fouling the bay, the added flushing of a new inlet can mask its effects.

As a result, the first reaction to the appearance of a new inlet has often been to demand that it be "stabilized" in place. By and large, this process has been disastrous for the nation's beaches, especially on the East and Gulf Coasts where barrier islands predominate. If the inlet channel is regularly dredged and the dredged sand is dumped into deep water, as is often the case, the effects are even worse. Some scientists estimate that in places like the east coast of Florida, where all but two of the nineteen inlets are artificially kept open, up to 80 percent of erosion is jetty- and dredging-related.

Sebastian Inlet, in Brevard County, Florida, is a classic inlet on a classic barrier island, about thirty miles south of Cape Canaveral. Randall Parkinson, a coastal sedimentologist at Florida Institute of Technology in Melbourne, is a consultant for the Sebastian Inlet tax district, and he often walks out to the middle of the bridge over the inlet, which offers a clear view of the coast from bay to island, north and south. From the bridge, he observed on one recent visit, it is clear that the ocean coastline is more or less straight. On the bay side, though, the shoreline is highly irregular, with bulges sticking out here and there. Each bulge, Parkinson explained, is the lobe of an old flood tide delta, "a sign of an old inlet that has built up with sediment."

Just inside the inlet the water loses its bright turquoise clarity. It darkens over the flood tidal delta, the lobe of sand Sebastian Inlet has hijacked from the littoral drift. "That flood shoal is a permanent sink," Parkinson said. "The sand is going to sit there forever unless someone dredges it out."

On the ocean side of the inlet, waves were breaking in an arc around its mouth, as if something in the channel were gently blowing them back toward the sea. Parkinson explained that the waves were breaking at the seaward edge of the ebb tidal delta, the thousand-foot shoal formed of sediment carried out of the inlet by the tides. The formation of the ebb tide shoal, the work of years, has put the inlet into equilibrium, Parkinson said. That is, relatively little sand is leaving the littoral drift to become tied up in the inlet or its shoals. Though Sebastian Inlet is jettied the shoal has become, in effect, a substitute path for sand moving along the coast.

"Ten years ago I felt all inlets were a disaster," Parkinson said as he looked out at the thinly broken surf. "And I wouldn't put any more in. But I have learned a lot. . . . I would tell you now this inlet is in equilibrium." Parkinson conceded, however, that this inlet is unusual. It is used by shallow-draft boats that can easily navigate its shoals. As a result, the ebb and flood tide deltas do not suffer constant dredging. This inlet can withstand the demands of people. Many others cannot.

Cape May, New Jersey, is a case in point. As long ago as 1801, Ellis Hughes, the postmaster of Cape May, was advertising it in the *Philadelphia Aurora*:

The subscriber has prepared himself for entertaining company who uses sea bathing, and he is accommodated with extensive house room with fish, oysters, crabs and good liquors. Care will be taken of gentlemen's horses. Carriages may be driven along the margin of the ocean for miles and the wheels will scarcely make an impression upon the sand. The slope of the shore is so regular that persons may wade a great distance. It is the most delightful spot that citizens can go in the hot season."[5]

By the time of the Civil War, Cape May had become one of the nation's most elegant beach resorts. By a quirk of geography, it lay south of the Mason-Dixon line, and it drew so many visitors from the South that at the outbreak of hostilities its loyalty to the Union was justifiably suspect. After the war, its resort business surged on, interrupted only briefly when a devastating fire destroyed much of the town in 1878. The rebuilding that followed the fire turned the city into a rare snapshot of Victorian domestic architecture that today draws thousands of admiring visitors. But the beach where vacationers once rode in their carriages, and where Henry Ford staged automobile races as late at 1908, is gone. In its place are a massive stone seawall and crashing waves.

"Our community is nearly financially insolvent," city fathers wrote later, in an appeal for federal aid. "The stone wall, one mile long, that we erected along the ocean front only five years ago has already begun to crumble from the pounding of the waves since there is little or no beach. . . . We have finally reached a point where we no longer have beaches to erode."[6]

What had happened?

Cape May had always been an unstable area, and as a result it was one of the first areas to undergo substantial stabilization with groins, artificial dunes, a seawall, and other structures. The major villain, however, lies to the north, at Cape May Inlet, which the Corps of Engineers cut in 1911 and stabilized with jetties. Eighty years later, Wildwood Crest, the town directly updrift, had the widest beach in New Jersey. Tourists in nearby North Wildwood complained so loudly about the half-mile walk from the Boardwalk to the water that the town planned a tram-car shuttle to take them there. And it decided to relieve parking problems by building a three-hundred-car lot on its "surplus" beach.

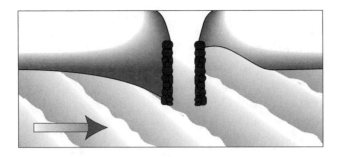

Once jetties stabilize an inlet, sand accumulates on the updrift side. Downdrift, erosion begins. *(Jen Christiansen)*

South of the jetties, though, the beach quickly shrank down to nothing. Cape May has been struggling with the problem ever since, trying everything from pumped replacement sand to artificial reefs. But its fine, natural beach is a thing of the past.

A similar drama played out on the islands south of Charleston, South Carolina, where the Corps of Engineers began building jetties in 1879. Some of the first effects were felt on Morris Island, immediately to the south, where a lighthouse had stood since 1767. The Confederate Army destroyed that structure in 1861, lest it serve as a beacon for Union warships, but a second lighthouse, formally the Charleston Lighthouse but known to South Carolinians as the Morris Island Light, was built on the same spot in 1876. It was 161 feet tall, with a base diameter of 33 feet, and it looked out over a wide beach backed with low sand dunes.

When the jetties were finished, erosion at Morris Island began to accelerate. Today, the island is a shadow of its former self, its coastline far inland from where it was in 1879, and the Morris Island Light stands derelict, half a mile at sea. In 1962, a lighthouse on Sullivan's Island, north of Charleston, took over the task of marking the entrance to the Charleston harbor.

Meanwhile, just as the jetty construction began, people began moving from Charleston to Folly Beach, the barrier island just south of Morris Island. Even as development was gathering speed, the jetties began to show their evil effects, and the beach at Folly Beach began to retreat. Today, after decades in which townspeople tried one remedy after another in a vain effort to save their beach, the federal government has stepped in with a multi-million-dollar beach-building effort.

In cities like Charleston, the costs and benefits of jetties are in a perverse kind of balance—losses in one developed area are countered by benefits to development elsewhere. This balance is thrown off when jetty construction damages public parks or wildlife refuges. Perhaps because all Americans have a stake in these lands, the feeling of loss is exaggerated. The parks south of Ocean City, Maryland, offer an example.

The development of Ocean City began in 1869, when a farmer named Isaac C. Coffin built the first "hotel" on Fenwick Island, a narrow barrier that ran forty-eight uninterrupted miles along the Maryland coast from Delaware to Virginia. His hostelry was little more than a cottage, but the good fishing on Fenwick Island soon began attracting paying guests who rode over from the mainland on a small ferry. Coffin named his establishment the Rhode Island Inn, choosing the name, legend has it, when a signboard from a ship by that name washed up on the beach in front of his house. Other hoteliers followed, especially after the island was linked to the mainland in 1880 by a railroad bridge. Eventually, a roadway from the mainland, electric lights, an amusement park, and a boardwalk came to the resort, which was named Ocean City.

In August 1933, a hurricane struck. It destroyed the railroad bridge, chewed up the boardwalk, sent cars floating out to sea, and wrecked several of the hotels. Overwash left streets buried in sand and cut the island in two, at the south end of the town. The storm was a disaster, but the inlet it left behind quickly turned into a blessing for the entrepreneurs of Ocean City. Now they could moor oceangoing fishing boats on the island's sheltered bay side and still get quickly to sea. The sport-fishing business bloomed. Commercial fishing grew, too. The inlet was more than compensating the damage the storm had done.

To stabilize the inlet, the Corps of Engineers built rock jetties along its edges and out to sea. Sand moving along the coast from the north, trapped by the northernmost jetty, widened the beach in front of the town. Now even bigger hotels were built, the boardwalk was improved, and a large arcade and amusement park took their place on the beach. There was so much sand near the jetties that the town paved much of it for a parking lot.

But it was a different story south of the inlet, on the barrier now known as Assateague Island. Currents continued to carry sand away to the south, but the jetties blocked the sand that would normally have arrived to replace it. Assateague Island began eroding fast. Today the stretch just downdrift of the jetties, the lush island the explorer Giovanni da Verrazano called "Arcadia," is duneless and treeless. Its only features are washover fans, places

A storm in 1933 cut an inlet at Ocean City, Md. Today, erosion has left Assateague Island vulnerable to storm waves. It has retreated so fast its ocean beach is now aligned with the sound side of Ocean City to the north. *(Stephen P. Leatherman)*

on the back of the island where storm waves have left sand deposits. The island, robbed of sand for decades, its dune defenses eaten away, is eroding furiously on its ocean side. Though Assateague and Ocean City were once neighbors on the same island, the ocean side of Assateague is now further inland than the bay side of Ocean City. Northern Assateague is retreating so rapidly that geologists predict it is only a matter of a few decades until it collapses into the mainland behind it and vanishes.

The remains of old sand fencing still trap sand, as do driftwood, shells, or even debris washing up on the beach. But the supply of sand is meager, and plants have a difficult time colonizing it. The most important of these plants is American beach grass, the dominant plant on beaches to the north, where its sand-trapping ability makes it a formidable dune-builder. But Assateague lies at the boundary of two major plant zones, so while there is a large variety of plant life there, many species do not thrive. Assateague is just about the southernmost part of the beach grass range, and the plant does not do well there. Its growth is further hampered on Assateague by feral ponies, beloved of tourists but despised by ecologists. These voracious horses crop plants down to the ground, destroying their ability to trap sand. Unfortunately, they love American beach grass. (Many vulnerable islands to the south owe a share of their survival to the fact that the ponies that inhabit them do not like sea oats, the plant that takes over from beach grass as the dominant plant on the dunes. But the ponies do plenty of damage, even there.)

Leatherman, who prefers the campgrounds of Assateague to the bright lights and honky-tonk of Ocean City, often took his students from the University of Maryland on field trips to the island. For years, the groups would stay at a sprawling house, clad in yellow shingles, that in years past had been the scene of elaborate hunting parties.

When the house was built, Leatherman said on one 1991 trip, it was two hundred feet from the first of several rows of dunes. By the time Leatherman began conducting field trips, however, only one meager dune separated the house from the waves. "A hurricane will take it out," he told the group, and he was almost right. A northeaster that struck weeks later, on Halloween, chewed the dune out from under the house and left it teetering on stilts, condemned. Heavy weather the next winter finished it.

Ironically, the jetty that is starving Assateague is not enough to keep Ocean City as a whole well fed, despite the fact that so much sand is piling up in the southern part of town that the jetty cannot hold any more. To the north, the shape of the ocean floor nearby, the flow of coastal currents, and other factors, imperfectly understood, have created what geologists call a "node"—an erosion hot spot. That means that the beaches in the center of town are in continual need of new sand. Coupled with the demolition of the island's original dunes (except for one small hillock fenced off on the beach as a reminder of its vanished glories), the erosion is a constant threat to the center of town.

Incredible as it may seem, fishing interests and government officials in North Carolina have been trying for decades to create the same miserable situation on the Outer Banks, at Oregon Inlet.

"Oregon Inlet has been a thorn in the side of a lot of people for as many as twenty-five to thirty years," says Robert Dolan, a coastal scientist at the University of Virginia who has consulted on government plans for the area. "It's probably one of the best examples we have in the coastal environment of a conflict between diametrically opposed mandates"—supporting a struggling fishing industry and preserving a pristine national seashore and wildlife refuge.

The Banks are at the far reaches of the North American continent, closer to the edge of the shelf than any place on the Atlantic Coast. Wave energy is high on this coast, and it is one of the most hurricane-prone areas in the United States. Winter northeasters routinely whack the banks with high winds and heavy surf that can last for days, reshaping the landscape with every high tide. The sand this heavy weather carries offshore has created a network of ever-shifting shoals whose treachery has made them, in legend and in fact, "the graveyard of the Atlantic."

Oregon Inlet opened just south of Roanoke Island late in the morning of September 7, 1846, in a fierce hurricane. It was named for a ship, the *Oregon*, that ran aground on its shoals soon after. Each year, geologists believe, currents carry seven hundred thousand cubic yards of sand south along this stretch of coast—the rough equivalent of a truckload every seven minutes around the clock. Some of it piles up on the north side of the inlet, in curving spits. Much of the rest ends up in the inlet itself, where it forms ebb and flood tide deltas and constantly shifting shoals. Left to themselves, these shoals might eventually clog the inlet so thoroughly that it would heal over. Instead, though, the inlet is constantly dredged, and the sand is disposed of as cheaply as possible—dumped in deep water offshore. As a result, the barrier south of the inlet, the Pea Island National Wildlife Refuge, is chronically, desperately short of sand. It is eroding fast.

The Corps of Engineers, at the request of local fishing interests and with the support, off and on, of federal and state officials, has proposed to end the need for dredging by building gigantic jetties—piles of rock two hundred feet wide at their underwater base, rising ten feet above the waves and extending almost a mile out to sea. This project would almost certainly turn the erosion of Pea Island into an even worse disaster.

Before the dredging began, Oregon Inlet followed inlet rules. Sand would pile up on the north side, as the spit grew there; shoals would form in the middle of the inlet; and, as a result, the supply of sand to the south was slowed. As the north side grew, the south side eroded. As a result, the inlet opening has been migrating south from the time it formed. A lighthouse built on the south side shortly after the inlet formed fell into the sea soon after. A replacement met the same fate. Finally, Bodie Island Light, one of the famous beacons of the Outer Banks, was built on the north side in 1872. It too has been a victim of the inlet's travels. Today, it is a mile inland, unusable.

When dredging began, barges carried steam shovels into the inlet channel and the great machines dropped their "clamshell" jaws down to scoop up sand. Erosion accelerated a bit, but this dredging method is relatively inefficient, and its effects were not too dramatic. In fact, they were slight enough that fishermen complained about the continuing shoaling problem in the inlet. So in 1982 the dredgers brought in hopper dredges, which are much more effective. Erosion to the south, in the Pea Island National Wildlife Refuge, accelerated dramatically. There are few structures there to serve as benchmarks for erosion, so until recently few people except the staff at the Coast Guard station south of the inlet knew how bad it was. They were watching the beach vanish before their eyes. (A particularly stormy winter took five hundred feet of beach, one of them said a few years later.) Eventually, the station was abandoned. A replacement was built on Pamlico Sound, north of the inlet. Now boaters complain it takes too long for rescue boats to traverse the inlet and reach the ocean, a situation that will only get worse if the inlet continues to move south.

Things would be improved a lot if the sand dredged out of the inlet were deposited on the beaches to the south, or anyway close enough to shore for it to rejoin the littoral currents and continue its journey. But this approach has been rejected. It would be too difficult and expensive to send sand barges into the rough surf close to shore.

The situation at Oregon Inlet is complicated by the need to preserve the highway bridge that runs across it, the only road link between the southern Banks and the mainland. Oregon Inlet, fast-moving and highly unstable, was an inauspicious place to put a bridge in the first place, but the desire to lure more tourists to the Outer Banks led the state to start building one anyway in 1962. From the beginning, problems were apparent. For example, by the time the Herbert C. Bonner Bridge was finished, a catwalk for fishermen on its north end was already useless. The water that had once been under it had filled in with a broad expanse of migrating sand. Meanwhile, erosion so threatens the bridge footings on the south side of the inlet that they had to be reinforced by a huge wall of rocks. The state is reconciled to spending at least $100 million on a new bridge,[7] but many North Carolinians fear that if some solution is not found for the inlet erosion even a new bridge itself will "go to sea."

Some in North Carolina fear an even deadlier problem. As the sand spit on the north grows inexorably southward toward the pile of rocks, the inlet is narrowing. If a big storm overwashes the barrier, the narrowed inlet may not be able to accommodate the water's ebb flow. As a result, Pea Island, already weakened by decades of erosion, could breach, creating a new inlet—and an economic disaster for the Banks.

Meanwhile, opponents of armoring worry that the rock reinforcement is the first step toward building the jetties, one of the most hotly debated projects planned for the coast. Pilkey, whose home base at Duke University has made him a major figure in the state's debates about the coast, describes himself as "almost out of control with anger" over the mere idea of building jetties there. But the jetty proposal is highly popular among the operators of scores of commercial and sport-fishing boats in towns like Manteo, in the center of the Banks. For them, Oregon Inlet would be a quick, easy, and cheap way to and from the Atlantic fishing grounds, if only its constantly shifting shoals did not make the trip so hair-raising. When the wind is up, waves break in the very throat of the inlet. The hopper dredges have improved things a bit, they say, but efforts to turn Manteo into a bustling commercial fishing port will never succeed until the inlet is jettied.

The jetty project has been delayed by almost thirty years of vicious argument. Some of the disputes involve the project's design, but, at bottom, the whole question hangs on effects the jetties will have on Pea Island and portions of the Cape Hatteras National Seashore further south. The project's backers scented victory in the Bush administration, when Interior Secretary Manuel Lujan suggested he would give it a go-ahead. His counterpart in the Clinton administration, Bruce Babbitt, pulled back. But state and local officials have refused to give up the fight.

The project's backers argue that erosion to the south would not be significant, especially if the project includes machinery to pump sand from one side of the inlet to the other. For many experts, these arguments make no sense. In a scathing report in 1987, an expert committee headed by Inman of Scripps Oceanographic Institution reviewed the corps's proposal and said that in estimating the movement of sand, a critical factor, the corps "disregarded some of the most basic principles of the accepted methods." Assertions that building the jetties would stabilize the inlet are "contrary to the engineering and scientific evidence" and "based on inadequate data that have been incorrectly interpreted." The report also said the corps had wrongly neglected to consider the possible effects of sea level rise and had given insufficient thought to keeping the channel open and preventing erosion by continuing to dredge but dumping the sand on land.

On top of that, it is far from clear that the giant pump system required to allow sand to bypass the jetties could ever be operated successfully. This pumping system would have to function in an unsheltered inlet subject to the highest wave energy on the Atlantic coast south of Maine. This kind of bypassing "has never before been successfully accomplished anywhere in the world," the panel said.

Experts on fish offered other objections. They said cost-benefit analyses of the project relied on unrealistically high catch estimates, given the depletion of the Atlantic fishery. If indeed the optimistic catch estimates became reality, these experts said, there would be a devastating crash in the fish population of the nearby Atlantic. Depletion of fishing stock is already such a serious problem there that the federal government has established fishing quotas, and the quotas are severe enough that many North Carolina fishermen are taking their boats to other states, or even other countries. Further, the experts say, the jetties would extend so far to sea that they would prevent fish larvae that move in the water near the shore from passing through the inlet to their nursery in the estuary behind it, reducing the fish population even more.

Pilkey, never a voice of moderation when it comes to coastal armor, is enraged that anyone would even consider such a proposal. In his view, the jetty plan is nothing more than collusion between the Corps of Engineers and the fishing business. Because the corps's budget depends on projects like this one, he charged, it has to back them in order to stay in business. While the 1987 report criticized the corps's engineering, he attacks its integrity. "I don't think they're just incompetent," he says of the officials in the corps's Wilmington, North Carolina district office. "To me and to many others this represents a watershed in corps duplicity, dishonesty, and incompetence."

He acknowledges that unless something is done even the enormous rock pile now guarding the southern approach to the Bonner Bridge will not be enough to save the span. Unless the rocks themselves are reinforced, they will be undermined and eventually fail, taking the bridge with them. But unlike virtually everyone else on the banks, Pilkey views that prospect with equanimity. He says adequate access to the refuge, the park, and the small Banks communities south of the inlet could be maintained with military-type pontoon bridging or by ferries.

Advocates of the jetties laugh at these ideas. It would be impossible to maintain a pontoon bridge across the wide, angry inlet, they say, and anyway such a bridge would block boats trying to enter or leave the inlet. Ferries would be in constant danger of running aground on constantly shifting shoals and, if they avoided them, would taken an hour or more to cross the inlet now crossed in under five minutes by bridge and causeway. Besides, they argue, ferries or pontoon bridging could not possibly handle the traffic the bridge experiences every summer day.

As long as the jetty question remains unsettled, dredging continues in the inlet. And while the dredging continues, erosion continues in the wildlife refuge. The barrier is narrow here, and often there is little more than two hundred yards between ocean and sound. The major question now is how to preserve Route 12, the state highway that runs the length of the bank. In many places, the barrier is so narrow, low, and featureless that the road routinely washes out, even in minor storms. In several places, the state has already moved the road away from the beach, but in the most vulnerable places there is no longer any room to move except into the marsh, which might violate wetland protection laws. Another possibility, elevating the road, would be dauntingly expensive, especially given the relatively small number of people who live in the area to be served.

The argument over Oregon Inlet has been raging for more than three decades. If people cannot agree on a solution soon, Nature will decide for them.

More than any other state, Florida pays attention to the problems with inlets. It has to—it has thirty-seven of them, and almost all are causing trouble. On its Atlantic coast, seventeen of the nineteen inlets were cut and reinforced to benefit shipping, fishing, or other interests. And Florida's Atlantic beaches have been paying the price ever since.

For example, an inlet was cut in 1951 and jettied soon after at the south end of Cape Canaveral, for the benefit of Port Canaveral, which today accommodates a submarine base and a growing cruise trade. But Brevard County beaches downdrift have lost one hundred yards of beach—or more. Some engineers estimate that since the creation of the inlet beaches downdrift have been deprived of more than fourteen million cubic yards of sand. Up to ten feet of beach are lost each year. Residents and officials say the county cannot afford the losses; 40 percent of Brevard County's property tax base is on the barrier island, and its beach-related tourism businesses pumped $430 million into the local economy in 1990.[8]

Elected and appointed county officials have asked the federal and state governments for help. Some want the government to pump more sand onto their beaches. Others want a bypassing system.[9] Hundreds of property owners have sued, seeking to have the Corps of Engineers and its jetties declared responsible for the problem. They hope they can force the corps to act quickly, or that they can win a settlement large enough to pay for beach replenishment themselves. Meanwhile, another Corps of Engineers study, in 1996, concluded that it would take almost six million cubic yards of sand—and at least $62 million over four years—to widen a twenty-mile strip of county beaches an average of fifty feet.

Robert G. Dean[10] of the University of Florida, one of the nation's leading coastal engineers, has estimated that as much as 85 percent of erosion in Florida is inlet-related. Without inlets, he believes, Florida might be relatively untroubled by erosion, especially on its Atlantic coast, where coral reefs provide some protection from heavy waves and the Bahamas function as a kind of breakwater, dulling the edge of wind and waves moving in from the east. As it is, Dean says, Florida spends $30 to $50 million a year to replace sand lost to inlet-related erosion.

Until very recently, most coastal engineers believed that the downdrift effects of inlet jetties were felt only relatively close to the inlet. Now however, more and more coastal scientists are seeing effects at great distances. For example, some engineers now believe sand deficits caused by the presence of jetties at Port Canaveral are felt as much as twenty-six miles downdrift.

The amount of sand lost varies according to how much is moving along the coast at a given spot. On the northern part of the Florida coast, Dean said, as much as 600,000 cubic yards may flow past—or into—an inlet each year. Approaching Miami, at Haulover, the figure drops to 10,000 cubic yards. Over the last five decades, Dean said recently, inlet dredging operations have taken fifty-five million cubic yards of sand out of these inlets and dumped it offshore. Applied to the state's beaches, he estimates, it would be enough to build the beach seaward by more than twenty-four feet. At today's prices, this sand is worth $10 a cubic yard or more—at least $550,000,000.

Each year in the United States dredgers take four hundred million cubic yards of sediment from channels and harbors to keep them navigable, and most of it is dumped at sea.[11] As many a coastal geologist has noted, this practice is the moral equivalent of cutting old-growth forest and burning it on the spot. But often there is little choice. If people and business rely on an inlet, it must be kept open.

One obvious response would be simply to dump the sand near the beach on the downdrift side of the inlet, where currents could pick it up just as if it had never been trapped in an inlet. The Corps of Engineers is willing to consider this step and, for example, has agreed to such a plan at Brookings, Oregon, where harbor jetties built in 1960 have produced considerable erosion downdrift. The corps plans to dispose of dredged material in a large underwater berm, or mound, just off the Brookings beach, in hopes that waves and swells will carry the sediment up onto the beach.

In many places, though, it is dangerous to operate a dredge near the beach, where surf is rougher and the risk of running aground is great. Instead, dredging companies simply take their barges offshore to where the water is deep and dump the sand there. If a company could be persuaded to dump inshore, it would surely charge more, and most government agencies must look for the cheapest way to do their jobs. Anyway, though the corps is cognizant of the problem and has been trying to minimize it where possible, when it awards a contract to dredge an inlet, its aim is to clear the inlet, not preserve the beach.

Also, sand that has accumulated in inlets often becomes unfit for the beach. It may be contaminated with heavy metals or other pollutants in the bay water. More likely, it simply picks up too much clay or mud carried into the bay or lagoon by inland streams. These fine-grained materials erode quickly if they are placed on a beach, and if sand contains more than 15 or 20 percent of "fines" it is judged unsuitable for beach use. Fine-grain material also tends to look dark and unappealing on a beach.

Finally, once sand has settled in a flood tidal shoal, it may become colonized by ecologically valuable plants or animals, sometimes even endangered species. At this point, many people would argue against disturbing this substrate, even if it were beach-quality sand that could be retrieved at a reasonable cost.

Now that there is widespread recognition that inlet dredging is bad for downdrift beaches, coastal engineers are thinking about ways to lessen its effects. This was the topic at a recent conference in Cape Canaveral. Before the inlet was cut, local officials in Brevard County maintain, the beaches were actually accreting—that is, they were growing wider. Since then, Dean estimates, seven million cubic yards of sand have been dredged from the inlet, and beaches to the south have been eroding accordingly. The federal government replenished them in 1974, but the sand did not last long. Another project began in 1994, but the sand began eroding as soon as it was pumped onto the beach.

The conference was attended by local officials and representatives of the area's tourism industry, whose fortunes wax and wane with their beach. Kevin Bodge, a coastal engineer with Olsen Associates, a Florida firm, suggested that one approach would be to give up the efficiency of the modern hopper dredges in favor of the old-fashioned clamshell dredges. "Hopper dredges are such efficient channel sweepers that they mix good and bad material to the point it's not usable on beaches," he said. By contrast, he said, "old-fashioned clamshell dredges can just pick up the good deposits." Then, they could be returned to the littoral drift or applied directly to the beach.

Parkinson has offered another approach: sand traps. If a deep spot were dredged in the inlet, he explained, water passing over it would slow, causing sand to drop out and concentrate there. Periodically, the trap could be cleared of sand, which presumably would be uncontaminated and ready for the beach. Both of these tactics, of course, depend on the dredgers' willingness to dump the sand on or near the shore.

Given the difficulties imposed by dredging, many people whose income depends on stable inlets—or stable beaches—are pinning their hopes on sand-bypassing, the process in which sand is mechanically carried from one side of a jettied inlet to the other. Machines pick up sand that has accumulated on the updrift side of the jetty and mix it with water to form a slurry. This slurry is then pumped through pipes or giant hoses to the other side of the inlet, where it is sprayed onto the beach or into the surfzone, whence it moves onto the beach. In this way, machines accomplish what nature would do, left to itself.

But nature is far better at moving sand than even the most efficient sand bypassing system, and the seemingly simple idea has proved difficult

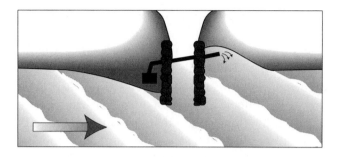

Bypassing can undo some jetty effects. Pumps carry sand, in a slurry of water, through pipes to the downdrift side, where it replenishes the beach. *(Jen Christiansen)*

and expensive to put into in practice. (Advocates of the Oregon Inlet jetties base their arguments in part on sand-bypassing, but the inlet presents a daunting challenge: moving seven hundred thousand cubic yards of sand a year.) If bypassing equipment is mounted on barges, rather than installed on land, operations would probably have to cease in winter weather. But some communities try to limit the use of bypassing equipment to the winter months, when there are fewer people on the beach. In either case, they must install and remove it every year, which adds to its costs.

The equipment itself is unsightly and noisy. If the pumps are powered by diesel engines they may pollute the air with smoke and fumes. A vigorous bypassing operation may render large stretches of sand unappealing or even unsafe to bathers. So it is hardly surprising that sand-bypassing plants operate successfully in only a very few places in the United States.

And even in areas where winds and waves are usually mild, maintaining equipment in the surfzone is difficult. Breakdowns occur regularly. For example, city officials have been trying for years to operate a bypassing system around the harbor at Oceanside, California, but the system has been plagued by failure. "It's a very simple, straightforward system and somehow the corps can't let a contract that will let it work," Inman complained. He attributed some of the problem to requirements that contracts go to low bidders, regardless of their experience.

The first-ever sand-bypassing system was installed in the 1930s at Boynton Inlet, cut in 1927 to improve water quality in Lake Worth, the lagoon inland of Palm Beach, Florida. Jetties were added in 1936. The following year, a bypass system began pumping seventy thousand cubic yards of sand annually from north to south. But it has not been enough. Communities to the south have repeatedly needed to replenish their beaches.

A bypassing system at Indian River Inlet, in Delaware, has been more successful. It began operating in 1990, when erosion threatened a highway north of the inlet. The inlet, which was constructed and jettied in 1938–40, interrupts the flow of sand that, overall, is to the north. The jetty south of the inlet has filled with sand, and more sand moving north tries to make its way around the jetties, only to be trapped in flood and ebb tidal shoals. By the mid-1950s, erosion north of the inlet had become so extreme that the state began nourishing the beaches there with sand removed from the flood tidal shoal or from back bay borrow sites. The beach there has been replenished every five years, more or less, with an average of 105,000 cubic yards of sand per year.

The system is designed to pump water from the inlet, mix it with sand trapped at the south jetty, pump the resulting slurry through a pipe running over the inlet bridge, and then spray it on the beach. When it is running smoothly, it is designed to deliver more than 200 cubic yards of sand an hour to the beach, enough to make up the 105,000 cubic yards previously added to the beach, on average, per year. So far, the system seems to have stabilized beaches north of the inlet, without increasing erosion on beaches to the south. Despite its success, the bypass system is not enough to ensure the health of the beaches north of the inlet. Periodically it must be renourished. And this system only needs to handle 100,000 cubic yards a year. Many inlets would require systems with much higher capacity.

In the 1920s, people tried to open an inlet on the Florida coast where Sebastian Inlet is today. "They used shovels," Parkinson related. "Of course, they failed." In the 1960s, however, dynamite accomplished what shovels could not. The inlet was blasted open, to a depth of twenty to twenty-five feet, and it was stabilized by the jetties. Today, many consider the Sebastian Inlet tax district a model of inlet management. For one thing, though the inlet lies on the boundary between Indian River and Brevard counties, it is administered by a taxing district including parts of both, whose boundaries reflect geography, not arbitrary political boundaries.

Also, the district aims to maintain the sand supply to the Sebastian Inlet State Recreation Area to its south. Already, one hundred thousand cubic yards of sand has been placed on the beach there, to make up for sand lost to the inlet, where it is estimated about half that much collects each year.

"Our task is to try to mechanically transfer the sand down to the beach," Parkinson says. In fact, the state of Florida requires inlet districts to bypass all the sand that would otherwise collect in inlets, though it does not have the enforcement resources to make the requirement stick.

But Sebastian Inlet is easily managed. The interests of both beachgoers and boaters are represented at the commission. "In nature, navigation projects and beach erosion are coupled," Kevin Bodge told the Cape Canaveral conference. "If these two interests are not coupled politically and legislatively, it is difficult to couple them any other way."

Also, at Sebastian Inlet none of the conflicting demands is unmeetable. There is little building at the park downdrift of the inlet. And the boats that use the inlet are small, so it doesn't take much dredging to satisfy their demands. "What you can see here is a perfect example of a manageable coastal situation," Parkinson said. "The inlet is not too big. The taxing district is relatively small. It is a manageable situation in terms of trying to meet all the needs of the user groups—the fishermen, the beachgoers, the home owners, the surfers, the turtles, the reefs, and the diving clubs."

But as he stands on the inlet bridge, he can see the march of houses and condos toward the coast. If development continues, he fears, the delicate balance will be destroyed.

CHAPTER FIVE

Unnatural Appetite

Everything goes somewhere.
—Ernest Callenbach, *Ecology: A Pocket Guide*

One sunny morning in the spring of 1970, Susan Miller set out to walk along the shore of Miami Beach, the barrier island across Biscayne Bay from Miami. It was low tide, and her plan was to walk the wet beach, the area under water at high tide but exposed when the tide is out. Years earlier, a federal judge had ruled that this intertidal zone belonged to the state of Florida, in trust for its people, and Miller, a reporter for the *Miami Herald*, wanted to see how it was faring.

She started out at the island's south end, at Government Cut, the artificial inlet dug in soon after the turn of the twentieth century to enhance the Miami harbor. There was a sandy beach here, part of a small city park. But when she headed north she immediately began encountering obstacles. There were bulkheads on the beach. Concrete seawalls and elaborate decks, some supporting swimming pools, extended into the water in front of the big hotels. Metal and wooden groins, long since decayed under the waves' assault, blocked her path. So much was built on the beach that there was hardly any place to walk, even at low tide. With the exception of another pocket-sized seaside park in the middle of the city, and a few isolated patches of sand here and there, her route resembled a military obstacle course more than a beach. The only dry way to traverse what once had been a beach was to clamber over concrete walls and trespass across zealously guarded hotel property. The alternative was to wade—or swim in deep water. There was hardly any beach at all.

By the time she reached the northern end of the island, Miller was in despair. Much of the beach had been destroyed, and she feared much of the rest would vanish in years to come. "I wrote in my notebook: 'How far could I walk seventeen years from now?'" she reported. "It wasn't a happy thought."[1]

But a little more than a decade later, Miami Beach had a beach again—an expanse of white sand almost two hundred feet wide, running from Government Cut north to Baker's Haulover Cut, the artificial inlet that forms the north end of the barrier island. Contractors working for the Army Corps of Engineers had built it, from sand they dredged from the ocean floor a mile or more offshore.

This beach, known officially as the Dade County Beach Erosion and Hurricane Protection Project, was not born easily, largely because the city's hotel operators bitterly opposed it. Their argument was simple: if the government paid to build a new beach, it would insist that the public have the right to use it. The hotel owners believed their guests would prefer a private pool overlooking the ocean to an actual ocean beach they would have to share with the public. Edwin B. Dean,[2] executive director of the Southern Florida Hotel and Motel Association, called the whole idea of building a new beach "a hoax" designed to seize beachfront property and destroy the privacy so prized by visitors to the resort. "Placing a two-hundred-foot-wide public beach in front of the hotels would create a major change in Miami Beach and its economy," he declared. "Miami Beach was built and became the world's number one beach resort because of private beach ownership which attracted hundreds of millions of dollars in risk capital."[3] If necessary, hotel and motel owners said, they would maintain their beaches themselves.

A few things eventually convinced them this approach was impractical. Edwin Dean began to change his mind, the story goes, when a regular winter guest canceled his penthouse reservation because the sound of waves slapping on the hotel seawall annoyed him. The Corps of Engineers produced a study showing that erosion was removing 161,000 cubic yards of sand a year from the 14.5 mile stretch of beach between Baker's Haulover and Government Cut. That was far more sand than the hoteliers could afford to replenish. The opening of Disney World in Orlando was providing stiff competition for tourist dollars. Finally, the lack of sand was starting to hurt business. Apartments and private homes away from the water were running down. Crime was on the rise. People called the resort "the beach that isn't."

Meanwhile, Miami Beach had a new mayor. He was Jay Dermer, a New York lawyer who had moved to Miami Beach in 1955 and run for mayor in 1967, on a pro-beach platform. Though his opponents called his plans "Dermer's Folly" and predicted a renourished beach would wash away in the first big storm, he defeated the incumbent mayor, FDR's son Elliott Roosevelt, and pushed ahead with his plan. Finally, in 1977, seven years almost to the day after Miller's ill-fated beach stroll, workers dredged their first bargeload of sand, mixed it with water to form a slurry, and began piping it onto the what would become the beach at Miami Beach.

Four years and $67 million dollars later, the project was complete. Though some sand quickly washed away—most of it fine-grained material that should probably not have been used in the first place—and though the project has needed occasional booster shots of sand over the years, the Miami Beach beach remains in place. Even Hurricane Andrew, which made landfall about twenty-five miles to the south, hardly damaged it. Despite the hoteliers' initial fears, the tourism economy has boomed. South Beach, the art deco neighborhood not far from Government Cut, has been restored into an architectural gem, its pastel low-rise hotels and apartment houses a favored background for television programs and fashion shoots. Further north, high-rise hotels and condos stand shoulder to shoulder along the sand, drawing tourists from all over the world.

Advocates of artificial beach replenishment cite Miami Beach again and again as evidence that beach renourishment is a practical approach to repairing eroded beaches. It shows, they say, that rebuilt beaches can offer lasting economic and recreational benefits. "It's a showcase project," says Robert Dean of the University of Florida.

But is it? And if it is, can its success be duplicated elsewhere?

The many problems with seawalls, groins, or other "hard" stabilization— and the absence of other acceptable alternatives—have left beach nourishment as the favored method of protecting development along the shore. Since 1970, scores of projects have been started up and down the Atlantic and Gulf coasts and in southern California, often as a last resort for communities facing severe erosion.

Some projects, like Miami Beach, have lasted a long time. But many more have washed away far more quickly than their designers anticipated. In a few cases, new beaches vanished even before contractors packed up the equipment they had used to build them. Other beach nourishment projects have ended up costing far more than original estimates. And while rebuilt beaches can bring economic benefits to some people, they exact substantial costs, monetary and otherwise.

As a result, arguments about the effectiveness, impact, and financing of beach nourishment have echoed loudly up and down the coast. Organizations like the Florida Shore and Beach Preservation Association, which organizes scientific conferences on beach erosion and lobbies for shoreline property owners, are fierce advocates of beach nourishment—and of government financing for replenishment projects. Other groups, like the Coast Alliance, a coalition of environmental organizations, call artificial beach-building the moral equivalent of throwing thousand-dollar bills into the surf. And all too often, these groups say, the dollars come from taxpayers who are paying to protect the second homes of people far more wealthy than they.

The National Research Council (NRC) attempted to settle the matter by convening an expert panel of engineers, coastal geologists, economists, and others to study the technology and its uses. The panel concluded in 1995 that beach nourishment was an effective method of shoreline protection for many areas, provided the projects are well designed and installed in areas where there is little underlying natural or human-induced erosion. But the panel acknowledged that not enough is known about why some projects live out their expected lifespans and others fail. In large part, it said, that is because little is done to monitor artificial beaches once the sand is pumped onto the shore. That, in turn, is because monitoring must take in not just the visible beach—what geologists call the "subaerial" beach—but also the underwater or "subaqueous" beach, which is difficult and expensive to track.

Engineers who design beach nourishment projects assume that much of the sand pumped onto the beach will end up underwater, but they cannot say in advance exactly how much will move, or how soon, particularly in areas where there are rough waves, severe storms, or offshore anomalies such as reefs, shoals, or rocky outcrops. "Although beach nourishment projects have been carried out actively for several decades, there is still not an adequate methodology to predict their detailed performance," the NRC panel said in its report. "This is due in part to the complicated alongshore and crossshore transport processes, the near uniqueness of every setting for such projects and the generally inadequate monitoring of both the forces on and responses of past projects to provide a basis for assessment of available methodologies and guidance for their improvement."[4]

In other words, until more is known about how sand moves on the beach, it will be impossible to accurately predict how long any given project will survive.

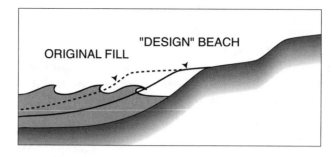

Schematic of a typical beach replenishment project. Because currents carry so much sand away so quickly, engineers initially place much more replenishment sand than needed to meet design criteria.
(Jen Christiansen)

In simplest terms, a renourishment project amounts to little more than applying a large amount of sand onto an existing beach. This infusion of sand disturbs the beach's equilibrium, and it adjusts by moving some of the newly applied sand underwater to maintain the normal slope of the beach. But this shifting of the sand occurs quickly, and it looks like erosion. Members of the public, whose enjoyment of the new beach has been unexpectedly cut short, are unlikely to be satisfied with talk about beach slope equilibrium. When they think about beach replenishment, they think of expanses of sand on which they can set out a towel and relax, not changes in the underwater portion of the beach profile. So advocates of beach nourishment projects routinely find themselves trying to explain what has happened when a beach they said would last for decades shrinks dramatically in a single stormy winter.

The renourishment process typically begins when a beach community experiences erosion severe enough to threaten shorefront buildings. Their owners pressure local governing bodies for action and they, in turn, approach the federal government for help, through the U.S. Army Corps of Engineers, which in addition to its work on coastal armor has become the leading organizer of beach nourishment projects in the United States.

The agencies involved begin studies to determine whether beach nourishment is feasible at the problem site, and they identify project boundaries, a decision fraught with political pressure. Then they must figure out how much new sand the site needs; allocate the project's cost (the federal government has typically paid 75 percent, though that figure has been reduced recently); find a source of beach-quality sand (a task getting increasingly difficult); and decide on how the sand will be picked up and delivered to the beach. The agencies, together with the private contractors who will do the work, then obtain the necessary permits and mobilize dredging and pumping equipment on the beach and at the sand source, or "borrow site." Then, the replenishment begins. It takes years, and sometimes even decades, to put the first grain of sand on the beach.

In most projects, sand is transported to the beach much as it was in Galveston nearly one hundred years ago, as water slurry pumped through pipes or heavy hoses and sprayed onto the beach. Depending on the size of the project, the pumping can take years, with pipes and pumps set up each spring and demobilized with the approach of winter, when severe weather can make the process dangerous or even impossible. Since World War II, there have been scores of major federal beach replenishment projects in the United States and hundreds of smaller, local efforts. More and more of them are planned as shoreline development accelerates and erosion continues.

The first beach replenishment project in the United States—and maybe anywhere—got under way in 1921 at Coney Island, in the New York City borough of Brooklyn. The project did not begin as a beach-building effort. Instead, its central aim was construction of a boardwalk to rival that of Atlantic City. City officials had nurtured the boardwalk idea since the turn of the century, but they were unable to achieve it, chiefly because of opposition from people who owned the property adjacent to the beach. Many of these people earned their living renting cabanas and other facilities to bathers, and they feared the boardwalk would encroach too much on the beach. A few others also pointed out that the Coney Island shoreline was always shifting, and questioned whether a boardwalk would be safe. An artificial beach answered both of these objections. With added sand the new promenade would not crowd out the bathers. Plus, it would buffer the natural changes of the shoreline. A series of groins was added to the plan to keep the new sand in place.

Between 1917 and 1921, the New York State legislature approved the measures necessary to allow the city to lay out a public beach about three miles long, from Seagate to Ocean Parkway, and acquire the land or easements needed to build it. Contracts were signed in October 1921, and work began the next year. As the groins were completed, workers began pumping sand dredged from a site about fifteen hundred feet offshore. They used one dredge, its pump powered by two steam boilers. Once the dredge was filled, the sand was mixed with water and pumped onto the beach through pipes running over the water on pontoons and then on wooden trestles on the growing beach. The project was halted briefly when the pipes suddenly began spitting out reddish building-type sand, not the white beach sand the contract called for. "This was a great disappointment and caused considerable unfavorable comment on the part of visitors to the beach," a municipal engineer, Philip P. Farley, wrote shortly after the project was finished.[5] Fortunately, the engineers found another source of white sand.

By the time work ended, in May 1923, an estimated 1.5 million cubic yards of sand had been pumped on shore, and the beach had been extended 330 feet seaward. The cost of the project was put at about $4 million, of which almost half went to compensate shorefront property owners. The city as a whole paid 65 percent; the rest was "a local assessment."[6]

"Though it was incidental to the construction of the boardwalk, the making of the beach turned out to be the most important part of the work," Farley wrote later in his report. Apparently, he said, "this is the first time that an attempt has been made to produce an artificial bathing beach by pumping from the sea to the ocean's edge and by pushing the high water mark seaward." He added: "It is of course too soon after completion to say with any degree of certainly what the ultimate result will be. As expected, the action of the sea has caused a general flattening of the beach with resulting changes in the position of high and low water lines. . . . Whether there has been an actual loss of sand . . . is quite difficult to say, since no re-computation of quantities has yet been made."[7] But sand seemed to be moving from east to west, he added, which would be consistent with previously observed patterns there.

Today, we know that this project was just the beginning of decades of work and tens of millions of dollars to stabilize, armor, and replace beaches at Coney Island and in the Rockaways nearby on the west end of Long Island. For New Yorkers, the effort has undoubtedly been worth it. These beaches are among the most heavily used in the United States and, as Farley noted, "the chief benefit that follows an improvement of this character is to be found in increased health and happiness of the vast crowd of people who are obliged to live in congested parts of a Great City and who have here an extensive playground for their recreation and enjoyment."[8]

But it is almost uncanny how questions and problems that bedeviled the Coney Island project persist today: objections from nearby property owners, problems finding beach-quality sand, difficulty keeping the sand in place, high cost, and uncertainties about why sand placed on the beach did not remain there.

Still, beach nourishment has been performed up and down the coast for decades. In Los Angeles County, almost all the beaches are artificial, built from sediment dredged out of harbors and sand removed inland in the regional building boom that began in World War II. Most Florida beaches also depend on regular replenishment for their survival. Delray Beach, to name one, has been replenished every few years since the 1970s, when waves began undermining revetments built to protect buildings as well as Route A1A, the major north-south roadway. The first major project in Delray Beach was undertaken in 1973; others followed in 1978, 1984, and 1992. On the whole, residents of the area consider the work a good investment. Also, though much of the sand has been lost from the project area, much of it ended up—at least temporarily—on nearby beaches.

To the north, Hilton Head, South Carolina, owes much of its appeal to artificial beach-building. Hilton Head is not a barrier island but rather a sea island, one of a series along the coast of South Carolina and Georgia. Though the soil on these islands is sandy, their shores are often muddy, perfect for shellfish but useless for sunbathing. When developers discovered Hilton Head in the 1950s, they immediately set to work, digging "canals" through the island to provide sand for new beaches. Today, these beaches, like those on other sea islands, must periodically be artificially replenished.

On the whole, replenished beaches do not have a very good record of longevity. Measured in terms of dry beach, not a single replenishment project on the Atlantic coast north of Florida has lasted more than five years, at least as viewed by Orrin Pilkey from his vantage point at Duke University. Pilkey has won a national reputation as the Jeremiah of the coast, and though he saves his sharpest barbs for hard stabilization like walls and groins, he crusades vigorously against irresponsible beach development and the beach renourishment on which it all too often relies. Single-handedly, he has inflamed the debate over beach nourishment, turning up the heat under the Corps of Engineers with charges of incompetence and deceit. For its part, the corps (led by James Houston, then director of its Coastal Engineering Research Center) fired back denunciations of Pilkey's research as sloppy and unscientific, based more on seat-of-the-pants observation than meticulous numerical analysis or computer modeling.

Both Pilkey and Houston agree that beach nourishment can offer substantial economic advantages to coastal communities. But Pilkey calls beach nourishment as practiced in the United States a "sham" or a "scam" because it is based on design practices that are unproved and untested by systematic monitoring. He says his studies show artificial beaches erode faster than natural beaches and that the mathematical models on which beach projects are designed are flawed at best.

Houston dismisses Pilkey's work as "very questionable because the underlying studies lack fundamental documentation that prevents others from examining or reproducing their results." In particular, he criticizes Pilkey for defining the "life" of a beachfill project as the time it takes for half of the newly applied sand to disappear from the dry beach. "High 'loss' rates that are unrelated to fill design are ensured when Pilkey and his colleagues primarily concentrate on the subaerial portion of fills and count 'loss' from the subaerial beach as soon as fill is placed," Houston wrote in one salvo in the battle the two carry on in scholarly journals. "The subaerial beach is only part of the beach system, and the entire nearshore profile down to an approximate closure depth has to be nourished."[9]

In a response to this critique, Pilkey asserted that Houston and the corps in general are beating the drum for beach nourishment at least in part to keep themselves employed, and cannot support their assertions that it is a cost effective approach to shoreline erosion.[10]

Recently, the corps has been adding economic arguments to its analysis of the merits of beach nourishment. Houston says the nation's investment in beach nourishment has been relatively small — 0.1 percent of its spending on crop subsidies or foreign aid — and its return in tourism revenue has been immense. The problem with this argument, though, is that it presumes that in the absence of nourishment there would be no beaches. This is not the case. In the absence of nourishment or armor, some buildings would surely be lost, but the beaches themselves would survive.

In any event, Pilkey himself is the first to confess that he relies little on mathematical analysis and computer modeling. He and his allies are contemptuous of numerical models that attempt to standardize beaches. In their view, even supercomputers cannot cope with the enormous variation in configuration, sediment supply, wave climate, and storm history of any one beach, much less all beaches. Though he was a member of the expert panel that produced the National Research Council report on beach nourishment, which touted advances in numerical modeling, Pilkey noted that the engineers themselves admit that when they compare their models to developments on an actual beach the models require "some adjustment of empirical coefficients to optimize the fit," as the report put it.[11] "Fudging," Pilkey calls it.

Citing the work of Dutch and Australian coastal engineers, Pilkey recommends close observation and detailed measuring of any beach for a decade or more before any replenishment project is begun. By then, it ought to be possible to calculate the loss rate of sand per year on that beach and, as a result, calculate how much sand the beach will need over a project's life span. Alternatively, he suggests what he calls "the kamikaze option"—simply dumping a lot of sand on the beach and waiting to see what happens. He notes that this approach has been widely though unconsciously used in the United States, particularly when sand dredged from inlets or harbors is disposed of on nearby beaches.

"This approach avoids the often heavy design costs common on U.S.A. replenished beach projects," he said. "Because beach design with mathematical models produces a beach of unknowable durability, the no-design approach may be just as successful." While beach projects built this way "may seem overly simplistic, they have many benefits: low cost . . . and greater intellectual honesty in that they do not fool the user into false confidence."[12]

The corps agrees that it is difficult to apply any particular remedy or formula across the board. "Most project sites have some unique characteristics and must be evaluated on the basis of their particular attributes in order to develop a project plan that affords the best balance between functional performance, cost-efficiency, return of economic benefits and environmental acceptability."[13]

And on the central point, Pilkey and the panel of experts are in accord: it is impossible to maintain a beach artificially at anything approaching a reasonable cost in an area of high erosion.

But a kind of Wonderland feeling begins to spread when geologists and engineers argue over individual beach nourishment projects. Folly Beach, South Carolina is a good example. Dean, of the University of Florida and also a member of the Research Council panel, calls it a success. Pilkey calls it a failure. Who is right?

Folly Beach is the island community whose development began about the turn of the century, just after construction began on the jetties at Charleston harbor to the north. Even as more and more Charlestonians were building their summer retreats or year-round homes at the beach, erosion accelerated. Storms in the 1930s and 1940s wiped out seventy-five feet of beach and extensively damaged dunes and beachfront houses. Between 1949 and 1961, forty-eight wooden groins were constructed into the surf, sticking out like fingers to trap passing sand. Later they were replaced with stone groins. But nothing worked. In succeeding years, winter storms and hurricanes took more and more beach, and by 1965, people were talking about a replenishment project. There followed more than two decades of studies, reports, new laws, and still more studies until, in 1986, the federal government agreed to pay 85 percent of the cost of rebuilding Folly Beach. (The percentage was relatively high, an acknowledgment that the harbor jetties, a federal navigation project, had contributed to the erosion problem.) A final round of engineering studies began — interrupted by Hurricane Hugo in 1989 — and in 1992 the town of Folly Beach began acquiring the shorefront property it would need to rebuild its beach. Much of this so-called property was already underwater. Nevertheless, the town had to purchase easements allowing it to put sand on it and allowing public access to the resulting beach. The easements were to run for fifty years, the "economic life of the project."

The replenishment effort, at an estimated fifty-year cost of about $116 million, involved moving 2.5 million cubic yards of sand from the Folly River and putting it along 5.2 miles of coastline. Some of the sand was formed into an artificial dune about nine feet high. Skeptics predicted it would be difficult to maintain the beach, especially the section near a Holiday Inn protected by a high concrete seawall, because of the high underlying erosion rate — five to ten feet a year. Very quickly, the beach disappeared in front of the Holiday Inn and officials in Folly Beach began accelerating the planned schedule of periodic sand infusions the beach already needs.

Another project, undertaken in the 1980s in Ocean City, Maryland, also ran into difficulties. Though Ocean City is trapping sand at the jetty at its southern end, an erosion "hot spot" at the center of town has been the target of several renourishment efforts. Like the project at Folly Beach, the Ocean City renourishment has been simultaneously derided as an obvious failure and praised as the city's savior.

The project was nearly completed in 1989, when a northeaster carried off much of the sand. Additional sand was applied, at additional cost, and in the summer of 1991 the project was completed and the equipment was removed from the beach. Almost immediately, devastating storms struck: the Halloween storm, a northeaster the next month, and, finally, the fierce northeaster of January 4, 1992. Almost everything that had been applied to the dry beach vanished into the waves.

Nevertheless, the Corps of Engineers and the project's advocates in the Maryland statehouse and in Ocean City counted it as a success. In the first place, the corps said, it could account for 95 percent of the vanished sand—it had moved offshore. Second, it said, the destruction of the beach only illustrates how much damage might have been done to buildings in Ocean City and how much it needs a "sacrificial beach" to protect its buildings from the sea.

"Certainly it saved a lot of destruction, but they got about half the protection they really needed," Steve Leatherman said after the January storm had clobbered the project. "But nobody wanted to tell the truth about the cost. In my opinion they needed $100 million. The corps is telling people it worked successfully, but there are no independent third-party evaluations. And it's running into real money."

Leatherman's criticism of the project hit a nerve with local and state officials, and he was summoned to the state capital to explain himself at a legislative hearing, where he told lawmakers he had a severe problem evaluating the project because he had been denied access to corps survey data. Having the corps monitor its own project was, he said, "like having the fox guard the hen house. . . . What is sorely needed is an independent, third-party evaluation and monitoring of the Ocean City Beach Nourishment Project." But anyone could tell, he added, that 80 to 90 percent of a beach project supposedly strong enough to survive a one-hundred-year storm had been "flattened" by a ten-to-fifteen-year event.

The officials' response was quick: they pointed out that they favored the project and that they were the ones who provided the money to pay Leatherman's salary and finance his laboratory at College Park at the University of Maryland. Leatherman subsequently left the university to head the International Hurricane Center in Miami.

It is too soon to say what will be the fate of the biggest project of them all, the huge effort at Sea Bright and Monmouth Beach, New Jersey. When the wall was completed, in 1926, it was fronted by a beach as much as 250 feet wide. (Maps from the 1880s show two small settlements seaward of where the wall is today.) But this part of the New Jersey coast is not fed by any sediment source, "except what was being eroded from the upland," says Norbert Psuty of Rutgers University. On this stretch of the New Jersey coast, upland begins right behind the beach and, as Psuty told a group of scientists on a field trip to the area in 1995, since the middle of the nineteenth century, when people began stabilizing the upland, "This area has been losing, losing, losing its beach."

Over the years, the loss of sand in front of the wall has steepened the natural underwater slope of the beach, producing ever bigger storm waves and still more sand loss. It is hard to imagine a worse place to try to build a beach. Nevertheless, after almost forty years of study and negotiation, that is what the Corps of Engineers is attempting to do. Plans call for a beach one hundred feet wide and ten feet above the mean low water line along the entire twelve-mile length of the wall. The sand sources are underwater ridges, relics of ancient dunes or barriers, two and a half miles offshore. Dredging boats pick up sand and then moor at a platform near the beach and pump the sand onshore. This new beach is only phase one of an even larger project that will eventually provide new sand for much of the New Jersey coast. The first segment, about twenty miles of coast, will cost at least $1.8 billion for its fifty-year lifespan, according to initial estimates.

Work began in Monmouth Beach in 1990, with the installation of more ten-ton boulders along the base of the failing wall. The first phase was budgeted at $250 million and involved ninety thousand tons of rock and twenty-three million cubic yards of sand. The federal government was to pay 65 percent of the project cost and officials at the Corps of Engineers estimated it would prevent, on average, $17.2 million in inland property damage each year.[14]

Four years later, pumps mounted on barges began spraying a slurry of water and sand into the waves where the beach would rise, and in 1995, the town had 350 feet of dry beach. True, much of it was closed to the public. In places, planners negotiating with abutting property owners won public access only to 15 feet above the high-water line. And true, the lack of public parking facilities made what public access there was theoretical rather than practical. Still, for the first time in decades, Monmouth Beach had summer traffic.

But a few voices were already being raised to question the corps's projection that the new beach could be maintained with booster shots of sand every six years. That estimate is "very optimistic," Psuty said, adding that every two years would be more like it. Otherwise, he said, "this beach will be back to the seawall." The long-term cost of the project will be far greater than original estimates if he and others are right, and already it looks as if they are. By the end of the summer of 1995, the corps was conceding that "unsettled weather" had resulted in the loss of much of the pumped sand. An additional 500,000 to 1.3 million cubic yards of sand was needed to replace it. Critics of the project, noting that weather at the shore is chronically "unsettled," declared that the missing sand proved the corps's six-year replenishment estimate had been wildly optimistic. A few weeks later, a northeaster, not particularly strong, had left some of the replenished area with only ten feet of dry beach. The corps attributed this problem to some sort of design flaw, but others suggested that the project should be stopped until engineers can figure out what the problem is.[15]

The corps maintained that the lost sand had either been redistributed along the shoreline, to the benefit of neighboring beaches, or had moved offshore and would return to the beach in calm weather. But Psuty and others pointed to the beach's steep slope underwater in front of the wall and predicted that much of the sand was probably in such deep water it would never be returned to shore.[16]

One stretch, in front of three condominiums in Monmouth Beach, proved impossible to replenish. For unknown reasons—maybe an offshore rock outcrop, maybe the configuration of the coast—sand on this stretch of beach eroded as quickly as it was applied. Corps officials began to whisper what others had been shouting: it may not be possible to rebuild a beach here. "Welcome to reality," said Beth Millemann, then executive director of the Coast Alliance, the umbrella group of environmentalists who oppose shoreline development.[17]

Before the project began, scientists studying the wall noted that it was settling, largely because wave action had scoured sand away from its base. The events of that summer confirmed critics' belief that a major task of the new beach, if it survives, will be to protect the wall.

Every beach nourishment project raises important questions and, all too often, planners and local authorities and members of the public do not pay them adequate attention. One of the most important is: Who benefits from this project? The question is especially important when the federal government is picking up most of the tab, as is typically the case.

The Corps of Engineers is required to analyze the ratio of benefits and costs of each proposed beach nourishment project before making the decision to go ahead. All sort of figures go into this calculation—predictions of increased tourism revenue, estimates of rises in property values and tax revenues, and calculations of how much of a town might be flooded in a coastal storm, in the absence of a broad beach.

But these calculations are not cut and dried. The mere discussion of beach-building can cause property values to rise, as prospective buyers anticipate the beach they will one day enjoy. Also, the very existence of a project can encourage more development in coastal areas. Though advocates of beach nourishment play down this problem, saying that beach-building projects are rarely carried out in areas that are not already thickly developed, critics counter that, in a vicious cycle, the projects encourage development whose safety hinges on future infusions of cash and sand to the beach that encourage still more development. Some critics of coastal development go so far as to assert that beach nourishment almost always ends up costing a community more than it generates in tax revenue.

The NRC report cited tourism income as a major benefit of beach nourishment. It conceded, however, that if people did not vacation at the beach they would probably vacation somewhere else. That is, revenue a beach project brings to a coastal resort is revenue that would otherwise have gone to tourism facilities in the mountains or elsewhere. (The exception, and it is a growing exception, is tourists from foreign countries who might have stayed home, or visited some other nation, had they not come to a beach in the United States.)

Cost-benefit figures are vulnerable to manipulation. So almost every announcement of a proposed project is accompanied by charges that someone has cooked the cost-benefit books, especially in communities where the value of the buildings to be protected is less than a proposed project's long-term costs. This situation is common in places where development is relatively thin and there are no major tourist facilities. "Atlantic City and Miami Beach—they'll renourish those beaches to the bitter end," Leatherman says. "It's worth it to throw sand out every three to five years." But there are many towns whose beaches are untouched by hotel, motel, restaurant, or casino—"below the line, in terms of economics," as Leatherman put it.

Faced with such statistics, some local officials may seek to economize by reducing the scope of a planned project. But this is an inefficient approach. Setting up a nourishment project is expensive. "The first sand grain to come out of a sand pipe costs $1 million," Leatherman says. "Your mobilizing-demobilizing cost is about $1 million. When you start to do beach nourishment, you have to do a pretty big project to make it make sense." Also, among coastal engineers, the rule of thumb is: the larger the project, the longer it lasts.

Calculating the actual cost of a beach replenishment project is further complicated by the fact that virtually every rebuilt beach will continue to need periodic infusions of sand. (Advocates call these beach nourishment projects "ongoing"; critics say they are "eternal.") These requirements are crucial to figuring out the cost of a project, but they are hard to estimate in advance. "There is a wide spectrum of costs," Tim Kana, a coastal engineer with a private consulting business in South Carolina told a conference on the subject at Hilton Head in 1993. "Some beaches lose half a cubic yard per foot of beach per year. Others lose twenty. Some sand costs $1 a yard, some costs $10. Costs can vary by three orders of magnitude." (When the Ocean City project began, the corps estimated it would cost $342 million to maintain the beach for fifty years, an estimate Pilkey dismissed as "propaganda." In its first five years, it soaked up more than $100 million.)

Kana often shows a cartoon to make this point in his presentations. The cartoon shows a municipal official talking to a coastal engineer about a proposed beach project. "How wide will it be?" the official asks. "How long will it last?" The engineer offers a baffling reply full of scientific jargon about tide lines, winter berms, wave climate, storm frequency, and so on. The dismayed official sighs. "What about artificial seaweed?" he asks.

Complicating things further is the fact that many beach nourishment projects offer temporary benefits—but to beaches outside the project area. This happens when sand washes off the replenished beach and begins moving along the coast to beaches downdrift. That is what happened when the state of South Carolina spent millions of dollars to renourish the beaches of Hunting Island, a state park about forty miles up the coast from Savannah. The project began to fail almost instantly, and some of the sand began to move south, to Fripp Island, where it created beaches in front of the rip-rap, seawalls, and revetments of the private, gated beach resort. The people of Fripp Island, who had already heard from Dr. Kana that further efforts to fortify their unstable island beaches would be in vain, were delighted. Without spending a penny, they were getting beach renourishment—though it was far from clear how long the sand would remain on their beaches. This was not the first time sand placed on Hunting Island had washed away to other beaches. In 1968, some 650,000 cubic yards of sand were placed on the beach, and almost all of it was gone in eighteen months. A second project was carried out in 1971, and it lasted less than half that long.

Charles A. Bookman, who as head of the Marine Board at the National Research Council helped organize the council's study and report on beach nourishment, spoke at the National Academy's 1992 retreat at Woods Hole on the problem of predicting project costs. "Engineers design a project and say it will last for ten years," he said. Then, as happened with Ocean City, "you get a storm and the sand goes somewhere other than at the high tide line and people say it's a failure. We need better ways of predicting design life, and more knowledge about where sand goes." Once projects are complete, he went on, "how do you know they have been successful? There is clearly a need to monitor. The sand will move. It may move offshore. It may not be lost to that littoral cell, but it may not be quite where you want it. There are some extraordinarily complex engineering issues here."

At first glance, the answers to Bookman's questions seem self-evident: if there was a beach where there once had been none, or a wide beach where before there had been only a narrow strip of sand, the project was a success. Indeed, that is the standard the average person applies. But when beach nourishment became the preferred method of shoreline stabilization—and many beaches seemed to vanish as soon as they were built—this criterion began to cause trouble.

Meanwhile, budget constraints had taken the federal government, and with it the Corps of Engineers, out of the business of building recreational beaches. So engineers were no longer able to count enhanced recreation among the benefits in their cost analyses. They began to talk of storm protection, a tacit acknowledgement that their work now is not to create a new or wider beach but rather to protect the houses, condos, restaurants, and shops behind it. This logic does not always work to residents' liking. For example, the corps figured out that a beach project to protect Panama City, Florida, from storm damage would require a dune so high it would have blocked the view of the water. Residents rejected the idea.

The political advantages of emphasizing storm protection, not beach building, become clear when storms occur and the new beach erodes far ahead of schedule. That is when designers speak of buildings protected, not the sand lost. They describe the project as "sacrificial"—designed to fail—and say the loss of the dry beach only means the project has done its job. In this framework, every project is a success no matter how long it lasts. Leatherman calls this way of thinking "success by definition—an absolutely stupid idea."

The construction of projects "designed to fail" raises problems for one federal program whose decisions can have vast financial implications for coastal property owners, the federal Flood Insurance Program. The program, run under the Federal Emergency Management Agency (FEMA), offers insurance at rates property owners would be unable to match on the open market—assuming they could find a company willing to insure them at all. But even this program must establish rates; rates depend on risk, and risk depends on the landscape. The problem is, how to judge the prospects of the replenished beach? How long will it be around to protect the property?

In setting premiums for property threatened by river flooding, the program takes improvements like dams and levees into account. Many advocates of shoreline development say the same practice should apply to beach nourishment projects on the coast. But these projects differ in an important respect: structures like dams and levees are designed to meet *and survive* a one-hundred-year storm. (The designation has been the source of some confusion on the coast—many people apparently believe that a storm of one-hundred-year magnitude can only occur once in a hundred years. In fact, the designation applies to the probability that a storm of that magnitude will occur in a given year. When meteorologists speak of a one-hundred-year storm, they mean weather that has only a 1 percent chance of occurring in a given year. But if a region experiences a one-hundred-year event one week, it could experience another one the next, and yet another the week after. Nature is unpredictable.) By contrast, beach nourishment projects are designed to be sacrificed to advancing storm waves. That means that when the storm passes, the artificial beach may be gone, leaving the insured structures—and the insurance program—at much higher risk.

There is a lively debate over whether the federal insurance program should "accredit" beach nourishment projects for rate-setting purposes. Insurance officials worry about declaring an artificial beach a permanent part of the landscape if there is no guarantee local, state, or federal governments will continue to cough up the money needed to maintain it. Also, what kind of stopgap protection should be required while a damaged beach awaits its next replenishment?

Todd Davison, who has studied this question for FEMA, points to Ocean City. "The flood came in, compromising the project. On the other hand, there were a lot of dry buildings behind the project that otherwise would have been wet." But until the beach is rebuilt, the area is far more vulnerable to storm damage than it was.

There are other towns, Davison continued, whose beaches depend on regular deposits of sand dredged to keep nearby inlets clear. "If the Corps of Engineers doesn't have the money to dredge or is delayed, there is a problem."

Many people with real estate or other property interests on the coast take the view that a sandy beach is infrastructure, sublime infrastructure, but infrastructure all the same. Just as a road, sewage plants, water lines, and the like must be maintained, they argue, beaches need regular repair and maintenance. Communities should provide funds to renourish their beaches just as they budget to repave their streets. Anthony P. Pratt, environmental program manager for the Delaware Department of Natural Resources and Environmental Control and a member of the NRC's expert panel, observed at one of the panel's meetings that few people in Delaware object to his state's spending millions of dollars each year to maintain roads to carry people to the beach. But replenishment projects are far more controversial. "We're used to paying for roads," he said. "We are not yet used to paying for beaches."

In any event, he continued, when it comes to evaluating the success of a nourishment project, the apt analogy may be tires, not roads. Motorists don't judge tires by how long they last, he noted, but rather by how many miles they travel. The equivalent for beach projects, he said, might not be how many years they last but rather the number and intensity of the storms they withstand. There are two problems with this argument: first, a beach, absent people and their buildings, would not need this kind of maintenance; second, road-builders and tire manufacturers have a much better record than beach-builders of estimating how long their projects will survive.

A few communities have suffered severe problems when they financed beach replenishment projects the way they finance other capital improvement projects—with bond issues, typically for ten-, twenty-, or thirty-year bonds. This practice fell into disfavor after many beach projects that had been predicted to last for decades failed within only a few years. Their failure left the community in the position of the home owner who buys cheap furniture on credit and must continue to pay for it long after it has broken or worn out. This argument was heard in 1996 in the coastal town of Pine Knoll Shores, North Carolina, after hurricanes chewed away at its dunes, leaving no beach at high tide. Town leaders proposed to spend $5.5 million to replenish their beach. Owners of oceanfront property would pay 88 percent of the cost—a beachfront home valued at $300,000 would have an additional yearly assessment of $1,749 for eight years. Opponents argued that a replenished beach would never survive a harsh northeaster or hurricane, but that they would be stuck with the bill anyway.[18]

The decline in federal spending has left coastal communities on the lookout for other ways to finance continuing beach nourishment. Some have added half a cent or so to their sales tax, others set aside some of the revenue from taxes on tourism facilities such as hotel rooms or restaurant meals. Property owners on Fire Island, on Long Island's south shore, moved to establish an erosion-control taxing district comprising the island's seventeen communities. In New Jersey, part of the fee the state makes owners pay when they sell property goes for beach replenishment. According to Ken Smith, a lobbyist for the state's shoreline communities, the fee program generates about $50 million a year, of which about $15 million is earmarked for shoreline "preservation"—usually beach nourishment after storm damage. "It's about half of what we need on an annual basis," Smith said, "but it's more than we had."

Critics of the federal government's role in beach renourishment say the people who benefit directly—owners of property nearest to the shore—should finance it entirely themselves. If they cannot afford it, put a lien on their property; its value has presumably been increased by the project. If they object to that, perhaps it's time to question whether the project should be done at all.

Even communities with enough money to finance beach nourishment are beginning to confront another problem: lack of sand. Since 1950, for example, New Jersey has used more than 40 million cubic yards of sand for beach renourishment; in the decade from 1985 to 1995, it used an average of 1.7 million cubic yards per year. Communities on the south shore of Long Island have used more than 80 million cubic yards since 1950. Large amounts of beach-quality sand are not necessarily lying around nearby waiting to be dredged up and applied to the beach. Unlike the beach, the ocean bottom is not uniformly sandy—much of it is muddy.

Still, on the East Coast, there are large amounts offshore in shoals or in the remains of ancient barrier islands or ridges, drowned thousands of years ago by rising sea waters. A law enacted in 1994 eased the procedures through which this sand is made available for beaches. The Minerals

Management Service (MMS), which among other things manages the off-shore resources of the United States, can now simply negotiate agreements rather than go through a long bidding process. Fees are assessed according to the amount and value of the resource—the sand—and the public interest served by using it.

Preliminary studies by the MMS suggest there are 400 billion cubic yards of sand in the mid-Atlantic shelf area, and 370 billion cubic yards in the south Atlantic. But the quality of much of it is unknown. Further, much of it is in deep water twenty miles or more offshore, which will make it dangerous—and expensive—to mine and transport to project sites.

If the sand source is too close to shore, however, mining it can make a bad erosion situation even worse. That is what happened in Grande Isle, the only inhabited barrier island on the Gulf Coast of Louisiana. A borrow site was chosen just offshore, but the hole the sand miners dug functioned as a sort of sand trap. Water passing over it slowed, and the sediment it was carrying precipitated out. As a result, a serious "erosion shadow" formed on the shoreline behind the site. Ultimately, the town had to armor the man-made erosion hot spot with rocks.

Accessible sand, like that trapped in inlets, estuaries, or harbors, may be contaminated with pollutants or even fine-grain materials like clay that are unsightly on a beach and erode much faster than ordinary beach sand. (In the New York City area, though, demand for sand is so great that commercial concerns treat contaminated sand and resell it.) A navy effort to pump nine million cubic yards of sand dredged from its San Diego facilities onto neighboring beaches foundered in 1997 when the sand turned out to contain mortar rounds and machine gun bullets.[19]

In some places, local authorities may disagree over who owns potentially usable sand deposits, such as shoals. For example, officials at the Assateague National Seashore and the city of Ocean City, Maryland, are looking at the same shoal as a source of sand. The shoal is forming through the action of the jetty at the south end of Ocean City, where sand began to pile up as soon as the inlet was stabilized in the 1930s. By 1950, the jetty had collected as much sand as it could hold, and new sand moving south was beginning to travel out to sea, where it formed a shoal at the inlet's mouth. By the 1980s, however, park officials and others interested in the fate of Assateague and the parks and wildlife it contains looked at the Ocean City shoal as a ray of hope. It had appeared to have reached an equilibrium size, and, at the same time, the rate of erosion on Assateague seemed to decline. Scientists concluded that sand was moving along the shoal to the island. Perhaps Assateague was not doomed after all.

By then, though, the beach nourishment efforts the city had underway to the north were already beginning to run into trouble. Always on the lookout for inexpensive sources of sand, city officials began eyeing the shoal at the south end, which is estimated to contain as much as eight million cubic yards of sand. The question is, Who owns this sand? "Assateague," answer the friends of the park. That is where it would go, were there no jetties at the inlet to stop it. "Ocean City," the city officials retort. We paid millions for it for our replenishment projects. Officials at Assateague are alarmed by this assertion of sand rights.

Florida is chronically short of easily accessible sand for beach replenishment, and two towns on its west coast, Venice and Siesta Key, almost ended up in court in the late 1980s and early 1990s, when Venice began looking at a deposit off Siesta Key for its own beach replenishment project. The state of Florida stepped in and claimed all offshore sand as a state resource and managed to steer Venice away from the sand. But as Florida's Deputy Environmental Protection Secretary Kirby Green told a conference later, Venice's rebuilt beach will need nourishment again early in the twenty-first century—when disputes will undoubtedly begin again.

Along parts of Collins Avenue, in the northern stretch of the Miami project, erosion is once again at work. It is no longer possible to walk on dry beach at high tide. Local officials in several Dade County towns organized replenishment efforts, but they were stalled by suits contending that the projects would damage coral reefs or turtle nesting sites. In some places, officials are hauling in sand from inland as a stopgap, and hoteliers are offering their guests free lunch to compensate for the lack of beach.[20] There, and elsewhere in southeast Florida, sand-starved communities are looking to the white sands of the Bahamas to replenish their beaches. Fisher Island, the private, 216-acre island enclave in Miami Harbor, built its beaches with sand imported from Bahamas sand, and the idea is becoming more popular—despite its cost and the nagging worry that there is something inappropriate about relatively flush Floridians negotiating with impoverished Bahamians to buy their very landscape. Bahamian tourism officials worry that by selling sand to maintain Florida's beaches, the islands are only encouraging Florida to draw some of their tourism dollars away. Environmentalists also fear that sand-hungry Americans and Bahamian officials, eager to make a deal, will sidestep important environmental considerations in sand-mining.

The Mineral Management Service must comply with environmental protection regulations in agreeing to sand removal. The agency is monitoring borrow sites and collecting data on how quickly plants and animals recolonize the site after the disruption of dredging for beach nourishment. So far, officials of the agency say, results are good. But there are still many who worry about how environmental standards will be set and monitored.

A few problems have already been uncovered. If a borrow site is near a coral reef, silt stirred up by the sand-removal process can cover the coral, killing it. The dredging equipment, especially the pipes that may run long underwater distances, can break or otherwise damage the delicate reefs. Once the sand has been dredged, new problems may arise. "This is the subaqueous version of strip and open bed mining," according to Rob Thieler, who earned his doctorate in Pilkey's program at Duke. "Holes in the seabed fill with mud. Diversity drops."

According to Psuty, dredgers working on the Monmouth Beach, N.J., replenishment project are attempting to avoid this kind of problem by "skim cutting" the offshore ridges they are mining for sand. That is, they are not digging holes but rather skimming sand off the surface of the ridges, lowering them by as much as twenty feet. It is unclear what effect this kind of dredging will have on bottom-dwelling creatures, but it may adversely affect the shoreline, Psuty says, because lowering the ridges by that much may increase the height of waves breaking on the shore, which would in turn increase its erosion.

Renourished beaches are rarely as satisfactory as the natural beaches they replace. They do not act, feel, or look like the beaches nature makes. Some of these differences are most severe at first and slowly vanish as the project ages. Others only get worse.

Ironically, beach nourishment projects themselves can be a source of erosion for neighboring areas. That is because a beach built out on the coast puts a bulge in the shoreline that can cause a so-called erosion shadow downdrift. This bulge of sand acts much like a groin, altering the longshore transport by trapping sand updrift and, in effect, starving the area downdrift. Coastal geologists call this kind of bulge, and its accompanying erosion shadow, an accretion wave. An exaggerated accretion wave can also form in the kind of ad hoc beach nourishment project in which an inlet is dredged and the spoil is more or less dumped on an adjacent beach. Normally, the bulge propagates down the coast depending on the background rate of longshore transport, taking its pros and cons with it.

Many people have noticed that replenished beaches do not feel the same as natural beaches. Often, they are far harder than natural beaches. There are several reasons for this problem, most of them related to differences between the sand nature put on the beach and that which has been pumped in the nourishment project. If there are too many "fines"—fine-grained particles—they can cause the sand to compact more easily, particularly when heavy equipment used in the replenishment project repeatedly drives over it. Also, material obtained from ebb tidal shoals or from the ocean floor may be richer in shell material than that of the natural beach. Depending on climate and the composition of the beach, the shell fragments can break down and react in unfortunate ways. "It can react with water and become indurated," said Parkinson of the Florida Institute of Technology, who has studied many replenished beaches in the state. "It interacts and you can get a natural cement," through a process called lithification or cementation.

Also, replenished beaches do not erode the way the natural beaches do. Wave action on renourished beaches often results not in the relatively smooth slope of a natural beach but in rather steep scarps cut into the edge of the sand. When the project at Monmouth Beach, New Jersey, began eroding, for example, waves cut a four- to eight-foot scarp along the shoreline. Children could jump down to the water's edge, but they often had a hard time returning to the dry beach. Steep scarps offer more than aesthetic problems. They can make it difficult for rescue vehicles to get to the water's edge, and they cause severe problems for animals, such as sea turtles, whose fate is intimately tied to the beach.

All species of sea turtles found in the United States are threatened or endangered. Biologists, geologists, and engineers argue fiercely about whether beach nourishment, overall, will help them survive or doom them to extinction. The argument that nourishment helps the turtles is simple: sea turtles nest on beaches, and if the beaches vanish there will be no place for them to nest. The argument against nourishment is more complex. The National Research Council reported in 1991 that beach nourishment projects can result in "significant reductions in nesting success" largely because the 250-pound turtles have difficulty climbing steep scarps or scraping the requisite two-foot nesting hole in compacted sand. Also, a nourishment project can bury nests under so much sand that hatchlings are unable to climb to the surface. Finally, it is unknown whether other changes in nourished beaches—differences in grain size or moisture content, for example—have any effect on hatchling survival rates, sex ratios, and so on.

The state of Florida, where most sea turtles in the U.S. nest, has proposed banning nourishment work for up to eight months a year, while turtles nest. The proposal would allow nourishment operations only in winter, when most operators usually shut down because of the danger of storms. Advocates of nourishment proposed a compromise, under which nourishment would continue in warm weather and beaches would be checked for new nests each morning. Any eggs laid overnight would be removed to hatcheries. But biologists pointed out that elaborate hatchery operations undertaken in Mexico have had little success so far in increasing the population of endangered turtles there.

For Parkinson, the sea turtle plays a role on Florida's beaches that the spotted owl played in the old growth forests of the Pacific Northwest—a surrogate for habitat protection. "People who have no great love for the turtles seize on the only tool they have to save beaches—the Endangered Species Act."

Many of the turtles' difficulties might be avoided, and the costs of beach nourishment overall might be reduced, with a technique that has been tested here and there in the United States for twenty years but that has never found wide use: pumping sand not on the dry beach but underwater just offshore.

A. W. Sam Smith, an Australian coastal engineer, has tried this technique with some success in his native Queensland, and he has been preaching it at conferences in the United States for years. Like his friend Pilkey, Smith condemns numerical models as useless. But, he says, if you study a beach long enough you will learn where it is most likely to build its own sandbar. Then, when it is short of sand, you can help it out by adding sand there. If you build up the dry beach, he argues, the sand will only wind up where the beach wants the bar, or underwater berm, to be. Plus, he adds, a berm can be built underwater for about half the cost of building one on the dry beach.

The Corps of Engineers has experimented with this idea in several places. Engineers for the corps considered building one at Folly Beach, but decided against it for fear it would ruin the surfing. But at one site off South Padre Island, Texas, approximately one hundred thousand cubic yards of sand dredged from Brazos Santiago Pass and the Brownsville Ship Channel, near the border of Mexico, was placed in a submerged berm about four thousand feet long and parallel to the shoreline in about twenty-six feet of water. The corps built another one, the National Berm

Demonstration Project, near Mobile, Alabama, where it weakened the wave energy hitting the beach.

The corp classifies berms as feeder, stable, or sacrificial, according to their function. Feeder berms are meant to migrate onshore or spread themselves out on the overall beach profile; stable berms are meant to stay in place, breaking wave energy; and sacrificial berms, like sacrificial beaches, are meant to hold off storm wave attack, even if they disintegrate in the process.

Cheryl Pollack, a corps engineer who described the berms at the conference at Hilton Head, said they should be treated as engineered structures. For example, if they are designed to stay in place and attenuate incoming wave energy (and trap sand), designers must anticipate accelerated erosion downdrift. Berms have been tested extensively in wave tanks, including the Large Wave Flume at Oregon State University, but few have been built in the field, so there are still many uncertainties about their ideal configuration, their survival, and their effects on nearby beaches.

Many coastal towns practice a kind of low-rent artificial beach-building, especially after storms, by sending bulldozers out on the beach at low tide to scrape sand and push it higher on the dry beach where, presumably, much of it would eventually migrate anyway as the beach recovered from the storm. Most coastal geologists do not consider beach scraping a serious contender in the world of shoreline preservation. They call it "Mickey Mouse engineering." But it does accomplish a few things: it builds a barrier, however small, against the next storm. And it gives people whose property is threatened the feeling that something is being done to help them—without the high environmental costs of other steps such as seawalls, revetments, or even sandbags. "Whether it does any good or not, it does make people feel better," says John Wells, a coastal geologist at the University of North Carolina at Chapel Hill, who has studied the technique. "But you're not adding any new sediment, you're just moving sediment that's already there."

Beach scraping is common in many states, particularly North Carolina, which has regulations barring hard stabilization on the coast. The state has set a number of restrictions on scraping as well. For example, the slope of the resulting grade must not endanger the public; that is, the scraper must not leave a trench in the beach. Many of the other restrictions aim to ensure that the sand scraped is not more than the beach could reasonably be expected to regain in a short period—a day, say. The maximum depth of scraping is one foot ("very frequently violated," Wells says). There must be no removal of sediment below the low tide line (also "very commonly violated").

Also, the scraping must not cause adverse effects on adjacent property. Is that a problem? "I don't know," Wells said. "It's an unanswered question. Basically, it's an experiment. You don't know how well it's going to work or what the impact is going to be until you do it. It's rather unsettling to know that we are taking a delicate system and performing experiments on it in hopes that it will do what we want."

Wells studied beach scraping on Topsail Island, North Carolina, a highly unstable barrier island that has nevertheless been highly developed. Wells compared two stretches of beach; one was regularly scraped, the other was not. After a large storm in the spring of 1989, he found that the dunes in the the area that had been scraped were somewhat more robust than those in the area that was untouched. Based on experience with Topsail Island, Wells offers some ground rules for beach scraping. First, recognize its limitations. Beach scraping offers some psychological benefits and even modest protection, but it will not prevent an inlet from forming in a vulnerable island. And if there is only a small distance between the structure you want to protect and the high tide line, there will not be enough sand to be of use. Finally, make sure that scraping does not harm the beach itself. For example, he said some of the problems Ocean City, Maryland, had with its replenishment project may have been due to overenthusiastic beach scraping.

Like much of what goes on at the beach, Wells said, "It looks simple, but it probably isn't."

In her book, *The Edge of the Sea*, Rachel Carson described a walk on the beach as "treading on the thin rooftops of an underground city." "Of the inhabitants themselves, little or nothing was visible," she went on. "There were the chimneys and stacks and ventilating pipes of underground dwellings, and various passages and runways leading down into darkness. There were little heaps of refuse that had been brought up to the surface as though in an attempt at some sort of civic sanitation. But the inhabitants remained hidden, dwelling silently in their dark, incomprehensible world."[21]

The strand Miami Beach built with such fanfare is not like that. The material placed on the beach at Miami was loaded with tiny shell fragments, and over the years it has solidified into a surface so rock-hard that shorefront entrepreneurs do a lively business renting beach chairs for people unwilling to rest on the so-called sand.[22] The hard sand is also inhospitable to beach dwellers such as ghost crabs, sand fleas, or other creatures. Even if they could dig their burrows here, they would find it hard to make a living. The seaweed and other beach detritus that forms a major part of their diet is deemed unsightly; hoteliers have the wrack line raked every morning, removing the little creatures' chief source of food. As a result, few of them dwell in or near the tidal zone and, in turn, the birds that feed on them look elsewhere for their meals. The result is a quiet beach, bereft of life.

Fortunately, this degradation seems to be little noticed by the vacationers who flock to Miami Beach, especially the fashionable folk who congregate at South Beach. A newly renovated hotel enjoyed a burst of publicity recently when it reopened, but the only clue that it was anywhere near an ocean was its South Beach address. The emphasis was entirely on its art deco ambience, fine food, and "water salon"—its pool. That approach is hardly novel. For many people who visit Miami Beach, the expanse of sand is for viewing, not for sunning or swimming. The quintessential beach experience in South Beach is to sit at an open-air table along Ocean Avenue and look out at the cars slowly cruising along the shore. The beach has become a backdrop—an artificial environment, artificially maintained.

CHAPTER SIX

Cause and Effect

By the law of nature, three things are common to mankind—
the air, running water, the sea and consequently the shores of the sea.
—*Institutes of Justinian*

In the middle of October 1993, hot Santa Ana winds began blowing over
the hills of southern California as they do each fall. But this year, the summer had been unusually hot and dry, so when brush fires started on the
hillsides just south of Los Angeles, the winds fanned the dry brush into in-
fernos. Firefighters struggled for days as the flames jumped from canyon
to sere canyon. By the time the weary crews had them under control, the
fires had reduced tens of thousands of acres to dust and ash

As those fires were dying in the hills above Laguna Beach, larger ones
were breaking out north of the city, in Topanga Canyon. From the air,
Los Angeles began to resemble a war zone, as Chinook helicopters
dumped three-thousand-gallon bucketfuls of water, scooped out of the
Pacific, onto the canyon flames and C-130 aircraft dropped bombs of fire
retardant through clouds of smoke so dark and wide they could be seen
from the space shuttle, 173 miles above the earth. These fires finally yielded to hundreds of firefighting crews, but miles of mountainside—more
than two hundred thousand acres—had become a dead zone, devoid of
vegetation except for the occasional tree, bereft of its leaves, standing in
bare, crumbling dirt.

For the firefighters, it was time to recover from a grueling marathon of
days and nights without food or sleep. For residents of the burned-over
areas, it was time to mourn their losses. But for Gregory Woodell, a planning specialist with the Los Angeles County Department of Beaches and
Harbors, it was time to make plans. "This fire is going to give us a lot of sil-
tation," he said, even as firefighters were catching their collective breath

for a final assault on the Topanga flames. "When we have the first rains this stuff is going to come down the creeks like crazy"—right onto the Los Angeles beaches.

Sure enough, the winter rains began a few weeks later, and mud began to slide. Without the plants that might have held the soaked earth in place, the hillsides began to erode, sending tons of sediment into newly swollen creeks. By early January 1993, mudslides had cut the Pacific Coast Highway between Laguna Beach and Los Angeles and again between the Los Angeles and Topanga. Still further to the north, residents of a tiny beach community, La Conchita, wondered if their town had a future after six hundred thousand tons of mud came roaring out of the hills, swamping the town. Sediment from miles inland was making its way to the coast.[1]

It was beach replenishment, California style.

Some people joke that California has four seasons: drought, fire, flood, and earthquake. And drought, fire, flood, and earthquake are the forces that have shaped the state's coast. For thousands of years, California's beaches were nourished in sporadic but enormous gulps when fire denuded the landscape, intense rain flooded it, and mudslides carried tons of sediment away. Hundreds of tons of dirt regularly thundered down the mountainsides, through the canyons and onto the beaches, where waves and currents sorted it and carried it along the coast. These floods of mud supplied the beaches with 75 to 95 percent of their sediment and created sandy beaches along much of the coast.[2] At the turn of the century, for example, a beach 100 to 250 feet wide ran forty or fifty miles from San Diego north. When people had to travel by wagon to Los Angeles, the beach was the quickest route.

But the European settlement of California, especially the twentieth-century development boom that took off in the "Southland" around Los Angeles, is a saga of increasingly elaborate efforts to bring nature to heel. In a process that began early in the century and accelerated after disastrous floods in 1938, the rivers that once fed sediment to the California coast have been dammed for flood control or to supply water to a growing population. As they trap water, the dams trap sand destined for the beach. By limiting floods, the dams reduce the kind of high-velocity water flow that picks dirt up and carries it along. Water flowing through the creekbeds lined with concrete cannot pick up any sediment to carry to the coast. When sediment trapped behind the dams is cleaned out, as it is periodically, it is usually used as fill on inland building lots. Still other canyons have been filled altogether, to make better building sites.

California's natural water flow has not been completely tamed by the state's continuing process of self-invention. But it has been channeled and weakened, and this manipulation of watercourses has almost eliminated the natural supply of beach sand. Never again, for example, will the Santa Clara River send more than eight million cubic yards of sediment to the beaches of Santa Barbara, as it did in one ferocious flood in 1938, the last to bring significant amounts of sediment to California beaches.[3] Spreading concrete has reduced the river's sediment supply by almost half.

By some estimates, more than one hundred million cubic yards of sand is locked away behind California's dams inland. According to Gary Griggs, the coastal scientist at the University of California, Santa Cruz, "In the past twenty years alone this coastal region has been starved of enough sand to build a beach three feet wide, three feet deep and sixty miles long. The Ventura River and Santa Maria River just to the north have had 66 percent of their drainages blocked by dams."[4] The Los Angeles River is contained in concrete for most of its 51 miles; 400 miles of tributaries are similarly armored. The Corps of Engineers and the Los Angeles County Department of Public Works have transformed these rivers into a gigantic system of storm drains running to the sea.[5]

Highway construction, starting with Highway 101 in 1912, also altered natural water flow, as drainage culverts were built to channel runoff.[6] Once highways were running along the coast, mudslides had to be stopped. So ever more hill areas were covered with concrete. If sliding mud did reach a coastal highway, it was cleared off at once—and almost always disposed of inland. Nowadays, even if people wanted to put it on the beach, strict regulations about what may be dumped there, and when, have created virtually insurmountable obstacles.

As if all that were not injury enough, over the years sand and gravel operators have mined hundreds of thousands of tons of sand from California beaches, dunes, and river beds themselves. In a final insult to the state's beaches, people eager to live on the coast cut off its last desperate sand source, the seacliffs that erode and collapse when a beach does not have enough sand. To prevent their clifftop houses from collapsing beneath them, residents have armored many of these bluffs with riprap, revetments, or seawalls. The beaches of California have been left to starve.

Douglas Inman of Scripps Institution of Oceanography had predicted years earlier that these beaches would gradually disappear as this sand supply was cut off. For a long time, though, Inman's prediction did not come true. The reason, ironically, was the very building boom that had caused much of the problem in the first place. California's growth was accompanied by heroic dredging efforts to build new harbors and marinas, and this dredging produced millions of cubic yards of sand that could be disposed of most easily through dumping on nearby beaches. The dredging boom began out of military necessity in World War II, when efforts to expand the navy base at San Diego produced, almost as an afterthought, a wide new beach for the Hotel del Coronado. Construction of Long Beach harbor, Marina Del Ray, the Los Angeles international airport, the city's enormous Hyperion sewage treatment plant, and other projects together contributed perhaps as much as fifty million cubic yards of sand to the beaches of southern California. When the Hyperion sewage plant was built in 1953, so much fill was dumped on the Dockweiler State Beach nearby that its underwater profile was sharply steepened, increasing the frequency of high, plunging waves to the point that municipal officials worried about the safety of bathers. Overall, more than one hundred million cubic yards of sand have been artificially applied to the California coast.[7]

Today, most of southern California's beaches are artificial, created as a by-product of these and other construction projects. But though occasional dredging efforts still take place and projects like the expansion of the sewage plant still cause Woodell's eyes to light up with thoughts of sand for Los Angeles County, the boom that built California's beaches has finally ended. Inman's prediction began coming true.

At first, the gradual disappearance of California beaches was masked by normal seasonal changes, as the beaches retreated under the onslaught of winter waves and advanced as gentle summer swells carried sand onshore. The problem is, less and less beach comes back every year, and if the winter is unusually rough, the deficit is dramatic.

The situation is complicated in places like Santa Monica Bay, where the normal flow of sand has been severely disrupted by an array of shoreline control devices. A municipal pier was built there in the 1930s, and the resulting disturbance in sand movement led eventually to the construction of five harbor breakwaters, three jetties, nineteen groins, five revetments, and other coastal "improvements."[8] Meanwhile, inland development cut off almost all the bay's natural sediment supply. The whole place is dead, as far as the movement of sediment is concerned, except when storms carry sand offshore. As a result, beaches that were once unnaturally wide are becoming unnaturally narrow. The underwater profile of much of the coast has steepened as it has been starved of sand. This deficiency, in turn, means that if sand is replaced on these beaches it will move offshore even more quickly than it would otherwise.

So it is no wonder that, as he looked out his office window in Marina Del Rey, toward the burning hills, Woodell could see a silver lining in the clouds of smoke. Already he was on the phone to officials at the Corps of Engineers, the local public works departments, and Caltrans, the state transportation department. He was hoping to use the inevitable mudslides to accomplish mechanically what nature had once done by itself.

Thirty miles or so to the north, in a house on Faria Beach in Ventura County, Katherine Stone was also pondering sources of beach sediment — and what she calls the politics of cause and effect. A lifelong Californian, Stone is a lawyer who specializes in environmental matters. She first became involved in coastal issues in the 1980s when the City of Oceanside, California, hired her to sue the federal government over a groin, built at the Camp Pendleton Marine base during World War II, that was blocking the flow of sand to city beaches. In the process, she met Inman at Scripps and other experts on the geology of beaches, as well as people involved in sand mining inland. Later, she and her colleagues combined Inman's theories on littoral cells with legal arguments about the public's rights to natural resources to devise a novel approach to protecting beaches.

If every beach is part of a natural system including sources of sediment, Stone reasoned, anything that interfered with the movement of sand to the beach was an attack on the beach. People who make that kind of attack should be obliged to "compensate" the beach, either by finding another way to carry the sand there or providing enough money to replenish the beach from another source. Stone called this new doctrine "sand rights."

It draws on legal principles that date from the sixth century, in the reign of the Roman emperor Justinian. Among other things, the *Institutes of Justinian* gave any Roman citizen the right to use shorelands or riverbanks to fish, tie up a boat, or unload cargo. This Roman law survived more or less intact in English common law after the Magna Carta and became the common law of the thirteen English colonies, surviving in turn as the law of those states. (The situation in the thirty-seven "new" states may vary slightly, depending on the legal definition of "navigable waterway" at the time they entered the Union.) Modified by the United States Constitution, the principle enunciated fifteen hundred years ago is known today as the Public Trust Doctrine.

Where once the British tidelands and waters were protected in the sovereign's name for the use and enjoyment of the people, today the doctrine protects "shorelands, bottomlands, tidelands, tidewaters, navigable freshwaters and the plant and animal life living in these waters"[9] for the use and enjoyment of all the people. Though private individuals can have a proprietary interest (*jus privatum*) in these resources, overall the state's interest (*jus publicum*) is inalienable. People or corporations may hold title to tidelands, shorelands, or other public trust property, but even if they are the only ones paying property taxes on it, and even if their deeds do not mention any other entity with an interest in the property, they own only the *jus privatum* interest and, in theory anyway, it is subordinate to the *jus publicum* interest of the public.

The public interest can be terminated but only narrowly, and only in pursuit of larger public purposes. For example, it can be terminated on parcels taken for construction of a wharf to advance navigation. In the eyes of the law, though, it is not possible for states to abandon, cede, sell, or give this *jus publicum* interest away, for the states cannot abdicate their obligations to their people. They cannot abrogate the trust, and no individual can claim a vested right against it. And though a state must not act outside its trust authority, when it takes steps to manage its public trust resources it does so with the rights of an owner, not the more tenuous authority of a regulator.

From time to time, issues of public trust are litigated. In 1987, for example, the Washington state supreme court ruled that a development company was not owed compensation when it was denied permission to fill tidelands it owned in northern Puget Sound, because the zoning that restricted it was in accord with the public trust interest in keeping the land natural.[10]

Historically, navigation, commerce, and fishing were the focus of the doctrine, and it was applied only narrowly to waterways, their beds, and tributaries. In recent years, however, courts have recognized a wider range of public trust lands and expanded their uses to include swimming, strolling, preservation of scenic beauty, and environmental protection.

In all twenty-three coastal states, this public trust gives the public at least some right of access to the intertidal zone, the "wet beach" between the low and high tide lines. (People in Massachusetts and Maine do not have the right to use this zone for recreational purposes, owing to unusual features in the charter of the Massachusetts Bay Colony, of which both were originally a part.) Courts in some states have also ruled that if people are to use this part of the beach, they must also have at least some access to the dry beach immediately above the high tide line. In a few places, states are even asserting that their right to protect public trust resources extends to barring activities that harm them. For example, they reason that if there is to be fishing, there must be fish, and hence regulation of pollutants that might harm them. Stone would apply this kind of reasoning to beaches. If there are to be beaches, they must have sand. Anything that interferes with their natural sand supply must be mitigated or halted.

The sand rights theory got a boost in 1989, when the Audubon Society won a suit against the city of Los Angeles over its diversion of water from the tributaries of Mono Lake, just east of Yosemite National Park. The city bought the rights to the water, but over the years the diversions caused the lake's water levels to drop lower and lower, destroying it as a nesting, breeding, or feeding place for birds and animals. In a path-breaking ruling, the Supreme Court of California ruled that this purchase did not give Los Angeles the right to take so much water that the lake's wildlife would be destroyed. Quoting Justinian, the court said the wildlife belonged to the state as a public trust and the government had an obligation to protect it. The city fought the ruling for years, but finally, in 1994, it agreed to stop depleting the lake's water supply and to act to restore lake water levels.[11]

"Putting natural law together with the facts that I had learned and Dr. Inman's theory of the river of sand and littoral cells, his theory of coastal processes, and learning what kinds of things interfere with that, I developed the theory of sand rights," Stone recalled years later. In effect, it is a public trust doctrine that, as Stone explains, "encompasses the movement of sand throughout the greater littoral cell, which is basically the watershed that feeds all the beaches—combining physical laws and laws of society."

If the shore is held in common for all, the reasoning went, the sand that feeds it must be held in common too. If that is so, anyone who restricts sand's movement to the beach is interfering with the rights of others. "The sand rights theory basically says that the coast is in motion and anybody who thinks they can build on the edge of it without causing something to happen is silly," Stone said. "When you build a massive flood control project to protect certain people's homes, not only are you helping them raise the value of their property, you are also impacting downstream on the beaches, which may be public beaches or may be beaches that are protecting somebody else's home. You're subsidizing one group over another. If you put some numbers on it, it's really astronomical. We have been basically subsidizing development on the coast with our system of law. The sand rights theory would readjust that."

Stone presented a paper on the subject at a conference in Santa Barbara in 1985; she and a colleague, Benjamin Kaufman, described the theory in 1988 in a paper in the journal *Shore and Beach*. Their paper described the problems that occur on the beach if sand is diverted, as well as some of the court cases suggesting that protecting the beach is a duty the states must perform. Their conclusion: the states must act to ensure that beaches get the sand flow nature had in mind for them. The states must protect the "sand rights" of beaches.

Stone backed her theory with several cases decided by the United States Supreme Court and by state courts, including a decision by the Mississippi Supreme Court extending public trust protection to activities other than navigation "such as bathing, swimming, recreation, fishing and mineral development."[12]

The California Supreme Court had made a similar ruling in 1971, and Stone cited that as well. That case involved a boundary dispute between two property owners on a bay in Marin County, and in the course of it the state court ruled that the public trust rights extend to environmental protection, recreation, and other activities.[13] The court said: "There is a growing public recognition that one of the most important public uses of the tidelands trust is the preservation of those lands in their natural state, so that they may serve as ecological units for scientific study, as open space, and as environments which provide food and habitat for bird and marine life, and which favorably affect the scenery and climate of the area."[14]

Theoretically, the prospect of being sued for substantial damages for interfering with the natural movement of sand should act as a kind of regulatory lever, forcing prospective developers to consider what their potential liability might be. So far, though, states have left it to aggrieved property owners or municipalities to force the matter. They, in turn, have had a mixed record in asserting anything like sand rights in court.

Florida actually has a kind of "sand rights" ordinance, a provision of legislation involving improved (jettied) inlets. The statute says that, on average, enough sand should be placed downdrift each year to compensate for the sand lost to the littoral drift. That is, sand trapped on the updrift side of the inlet must be moved to the downdrift side. Overall, this statute has been honored in the breach, as erosion problems up and down the coast of Florida attest.

In one celebrated case, decided in 1990, the town of Ocean Ridge, Florida, and others, sued the South Lake Worth Inlet District, saying that the jetties it constructed at Boynton Inlet, which was cut artificially in 1918, were disturbing the flow of sand along the coast. In effect, the plaintiffs argued, the inlet district was robbing Ocean Ridge, the town downdrift, of sand that would otherwise move onto its beaches. The state permit for the jetties required the district to move sand from the north side of the inlet, where it was piling up, to the south side, and the district established a bypassing plant with a capacity to transfer 65,000 to 70,000 cubic yards a year. After weighing expert testimony, the court decided that this bypassing was inadequate. The natural littoral drift was more like 185,000 to 200,000 cubic yards a year, it said, and the bypassing plant did not operate all the time.

The court ordered the inlet district and the county to renourish the beach south of the inlet for a distance of 2.5 miles, with at least a million cubic yards of beach-quality sand. Further, it directed the defendants to increase the capacity of the bypassing plant to maintain the shoreline established by the renourishment. The court also told the inlet district and the county to institute long-term monitoring of the bypassing and renourishment projects.

A similar situation arose in Easthampton, New York, on the south shore of Long Island, where three small groins are trapping sand moving along the coast. Ronald Lauder, the cosmetics magnate, feared erosion caused by the groins would eventually affect beachfront property he owns downdrift. Lauder hired Leatherman to investigate the situation and recommend steps to fix it.

In an earlier suit involving beachfront property immediately downdrift, the Corps of Engineers asserted that the erosion east of the groins was caused not by the groins but by a small inlet that is periodically cut to flush a coastal pond with salt water. The corps said sand that would have moved east was trapped in the pond's flood tide delta. Leatherman ridiculed that explanation. The entire pond would have to fill with sand to account for what was missing along the beach, he said. The groins, not the pond, were trapping sand.

Leatherman proposed a drastic solution: shorten the groins or remove them altogether. If the groins are shortened, Leatherman said, "Immediately you'll start seeing sand moving around them. It would only take a couple of years before each would be more in balance. In a couple of seasons you wouldn't see the shoreline being bumped in."

But this step proved impossible to accomplish. For one thing, it seemed clear that someone would have to give permission to alter the groins. But who? Federal, state, county, and local officials were unwilling to claim them, perhaps because they feared that whoever owned them would one day be forced to compensate others for the damage they caused. Also, there was the question of what to do with the boulders used to build the groins. Getting them out was almost certainly going to be more trouble than putting them in.

The sand rights issue also emerged in the argument over the shoal at Ocean City, Maryland. "We've taken a pretty firm stand on this," said Gordon Olsen, a resource management specialist at Assateague. "We'd prefer not to see that shoal disturbed. We feel that sand should have been coming to Assateague all along."

The Corps of Engineers acknowledges sand rights, in a way, in its formulas for determining how much it will contribute toward beach nourishment projects. The percentage increases—sometimes dramatically—the more it can be shown that a beach has eroded because of navigation improvements or other projects the corps has undertaken. Usually, these improvements involve jetties at inlets.

So far, though, sand rights remains a theory rather than an established legal doctrine, at least as far as the U.S. Supreme Court is concerned. The Court has yet to hear a case in which sand rights is the central question.

Though she is a lawyer, Stone says litigation is not necessarily the best way to make sure that beaches get the sand they need. (The client that drew her into the issue, the city of Oceanside, abandoned its plans to sue the federal government over the sand-blocking groin in favor of installing a bypassing system designed to move the sand past it.) And she does not insist that development be undone. Instead, Stone and other advocates of sand rights say, local and state governing authorities should take steps through taxes, fees, or regulation to require developers to assess the impact their projects would have on the beach and take steps to mitigate it.

Requiring people to fully mitigate or pay to mitigate the effects of their projects would be a strong disincentive to development that damages the beach. It would encourage people to keep away from the most vulnerable areas on the coast. It would also make a more accurate assessment of true costs of navigation works, waterpower, flood protection, irrigation, housing, and other development.

"The essence of 'Sand Rights' is a fair apportioning of the benefits and burdens of activities that interfere with the natural littoral process within a littoral cell which reduces sedimentation to our beaches," Stone wrote in 1989, in a paper presented at a meeting of the California Shore and Beach Preservation Association. "This can translate, for example, into impact or mitigation fees on new developments that contribute to the problem, including, in some cases, developments that are far outside of the coastal zone."[15]

Inman put it this way:

Dams provide water and intercept the normal flow of sand to beaches. Dams benefit cities, industry, and agriculture and damage beaches, coasts, and coastal communities. We must reconcile this inequity to coastal communities and beach users. The cost of nourishing beaches with the same amount of sand that is intercepted by dams would be a legitimate part of the cost of using water. The cost of replenishing beach sand should be borne by water users.[16]

Stone says government authorities could take any of three legal avenues to "protect" the sand rights of their beaches. First, state courts could recognize state responsibility for sand rights and insist they act on it; second, the legislature and Congress could require that designers of any project that might affect the beach take the effects into account and mitigate them; and third, public agencies could deal with the situation administratively.

For example, regulatory authorities could add sand supply to the list of factors considered when developers apply for necessary permits. "All environmental impact reports in coastal areas should have mandatory sections that analyze the effect of the project on the beach and mitigate to the fullest extent possible and consider all feasible alternatives," she says.

A simpler approach might be for local or state authorities to build in some dedicated source of money—development fees, say—that they could use to provide and retain sand for beaches. They could levy special taxes for beach preservation. Or they could set mitigation fees on activities that worsen erosion, or on development in general. Or they could modify existing user fees or permit fees—for sand mining, for example—to include a component for financing beach nourishment.

Inman advocates a step that derives more or less directly from his theoretical work on littoral cells: the formation of official agencies whose jurisdiction covers an entire littoral cell, with responsibilities for its beaches. The example he cites is a coastal area north of San Diego. The source of sand for the beach there is the hilly landscape inland; the cell's sediment transport paths are the rivers and creeks that carry sediment to the ocean and the currents that carry it along to its submarine canyon sink. This area is subject to a number of competing authorities—in this cell alone, the coastal towns of Oceanside, Carlsbad, Del Mar, and San Diego all claim jurisdiction. Inman was among those who successfully pushed the state legislature to pass joint powers legislation empowering coastal cities within a cell to form a beach district to act, as Inman put it, "as a unified body in trying to preserve and study and conserve their beaches."

A number of coastal municipalities in Santa Barbara and Ventura counties have established a district commission. The district commissioned a study of the Santa Clara River that discovered that the river used to bring about 600,000 cubic yards of sediment to the beach every year. Today, it carries only about 150,000 cubic yards. "The downstream communities were in much more serious long-term difficulties than they realized," Inman said.

But the district's authority is not as extensive as it might be, because the joint powers measure applies only to coastal towns, not to littoral cells as a whole. "It should include the whole area," Inman said. "It could. But it's difficult to get someone living up in Palomar Mountain, for example, to think that he ought to spend a lot of time worrying about the beach. But things that happen up there do have a direct impact on the beach."

As Inman sums it up: "It's a matter of trying to educate people so Kathy's sand rights legal arguments and my physical environmental arguments are all to be heard. . . . It's a very slow process."

A first step, some in California say, would be to relax environmental rules to expand opportunities to put sediment on beaches. At present, rules discourage or bar sand placement if material trucked from inland might not exactly match material already on the beach. Also, dumping sand may not be allowed if it might disturb endangered species, such as nesting shore birds.

"We have to look into changing the way we think about nourishing our beaches," says David Skelly, formerly of Scripps and now a consultant.

"Say a developer who has a hundred acres within a mile or two of the coast is grading some of these formations. There's significant source of sand in the material, but he can't put it on the beach. The permitting is a pain. It might have a higher percentage of fines. You have to put *exactly* what's natural to the beach on the beach. But when Mother Nature nourishes a beach it doesn't do that. Cliff failure—that's Mother Nature nourishing a beach."

He said problems with regulations keep a lot of potentially useful material off the beach. "We don't allow people who have something that would be reasonable to put it on the beach because somebody doesn't want to see a little more silt for a few days. But you place that stuff on the beach and let the waves filter it out or put it on a unused section of beach or do it in the winter when nobody's on the beach."

Instead, he went on, "They're putting it in inland landfills so that all the canyons will be flat so they can build on them. They end up trucking it a hell of a lot farther. We have overencumbered ourselves with regulatory commissions. If it doesn't offend one person, another person is going to go out of their way to find a way that it offends them. There's no incentive for someone to even want to go into this—even someone who might have the purest of motives."

Woodell, the Los Angeles County beach specialist, described his experience with this kind of problem, in a speech at a meeting of the California Shore and Beach Association in 1993. Several years earlier, he said, work to enlarge the major Los Angeles sewage plant made thousands of cubic yards of sand available, but city officials balked at giving him the permits he needed to put it on the beach. "I was getting stonewalled," he told the gathering. "They said, 'Drop it, you can't do it.'"

But Woodell was determined he would get the sand one way or another. ("It was either that or resign," he told the conference.) By chance, it was an election year, and one of the officials he was arguing with in the public works department was active in Mayor Tom Bradley's reelection campaign. So Woodell called the official and threatened to make the dispute public. To close the deal, he added mendaciously that he had already had two calls about the matter from the *Los Angeles Times*.

"He called back in five minutes," Woodell concluded. "He said, 'You have your sand.'"

On the morning of March 26, 1995, Interior Secretary Bruce Babbitt went to the Glen Canyon Dam in northern Arizona, in the Grand Canyon National Park, and took a tiny, temporary step toward restoring the Colorado River to its natural self. Pushing a button on the dam's control system, he unleashed a flood of water—forty-five thousand cubic feet per second—that came booming out of four outlets into the river six hundred feet below. Within an hour, the river began to take on its ancient copper color, as sediment trapped behind the dam entered the water. Slowly, this sand collected along riverside beach areas that had been starved by the dam. It was the first time water had flowed close to normally along the river since the dam was completed in 1963.

A major impetus for the flood was pressure from the tourist industry, particularly companies that organize rafting expeditions on the river. They need the beaches where their customers can rendezvous and relax. Also, archaeologists and preservationists argued that adding sediment to the river would reduce its capacity to erode important sites along its banks. But utility companies watched the process with alarm. They buy much of their power from the dam, and adjusting the water flow there to protect the downstream environment could cost them millions of dollars a year, they said.[17]

The flood was short-lived. After seven days, the authorities gradually reduced the water flow until it returned to preflood levels, a maximum of thirty thousand cubic feet per second. Still, its results were dramatic. Less than three months later, the Bureau of Land Management reported, there were more than fifty-five new beaches on the sixty-one-mile stretch downstream of the dam. (Nutrient-rich sediment, released from the riverbed, also brought a resurgence in fish and bird populations, the bureau said.)[18] The benefits led officials of the Department of the Interior to order that floods be held regularly in the future. They recognized that flooding is the kind of "disturbance" the river ecosystem needs to stay healthy.[19]

In the week the river flowed more naturally, 120 billion additional gallons of water went through the dam—enough to supply the city of Los Angeles for seven months.[20] Eventually, this water and the sediment it carried might have reached the Gulf of California, and who knows what might have happened on the beaches there. But the water is far too valuable to waste on a mere beach. The water, and the sediment it carried, was impounded again at Hoover Dam, downstream.[21]

CHAPTER SEVEN

The Big One

Even a low risk of a catastrophic event
must be avoided.
—Robert McNamara, *In Retrospect*

People who worry about hurricanes worry most about something they call
the nightmare storm, a big hurricane that would strike the East Coast at
Miami, head west across Florida to Fort Myers or Tampa, and then move
out across the Gulf of Mexico to make landfall again near New Orleans.
This storm would pass over some of the heaviest development in Florida
and through vulnerable, low-lying areas of Louisiana. Property worth billions of dollars would be destroyed, cities such as New Orleans would suffer devastating floods, and hundreds of people, many of them frail and el-
derly, would surely lose their lives. In midnight darkness of Sunday,
August 23, 1992, meteorologists at the National Hurricane Center in Coral
Gables, Florida, looked at their instruments and watched this nightmare
begin to unfold.

Only two days earlier, readings from radar, satellites, and reconnaissance aircraft had shown a weak tropical storm, whirling aimlessly near
the Lesser Antilles, so feeble its eye was barely discernible. As it moved toward the Bahamas, however, the storm gained energy from the warm
Caribbean water and gathered itself into a tighter and tighter cyclone, its
winds moving around its eye with ever greater speed. On Saturday morning, August 22, it was officially declared a hurricane, its winds over 75
miles per hour. Then, in thirty-six breathtaking hours, the winds had dou-
bled in strength, to almost 150 miles per hour, and Andrew, for that was its
name, was a strong Category 4 hurricane, verging on Category 5, the
fiercest. Worse, it was moving at more than 20 miles an hour, straight toward the developed heart of Florida's southeast coast. The meteorologists

at the National Hurricane Center estimated it would make landfall at Miami a few hours after midnight Monday.

The scientists at the hurricane center issued their first watch for Florida, covering the coast from the keys to Cape Canaveral, on Saturday. As is their practice, they divided the watch area into three zones. There was a one-in-five chance a hurricane would strike one of them in the next forty-eight hours.[1] By Sunday, the odds for each had risen to one-in-three and the watch was upgraded to a warning—the hurricane would strike within twelve to twenty-four hours.[2]

Less than a third of the people who live on Florida's southeast coast had left their homes for shelter inland by midday Sunday, but television weather forecasters in Miami were already telling those who remained that it was too late to escape. If they headed inland now, the meteorologists said, they risked the extreme danger of being caught on the highway when the hurricane struck. Anyway, the low-lying causeways leading off the barrier islands were already awash. "There is no safe place in South Florida tonight," Brian Norcross, the meteorologist at WTVJ, the NBC affiliate in Miami, told his viewers.

It was Norcross who, months earlier, had defied a chorus of skeptics at the station and insisted that it install emergency equipment for broadcasting in the event of a hurricane. Tonight, WTVJ's entire operation had moved into his makeshift storm center. As the wind rose, the rain poured down, and power failed throughout the region, the station remained on the air, broadcasting phone calls it received from frightened Floridians and the last-ditch advice Norcross offered them: retreat to your bathroom, where water pipes provide some reinforcement to the walls; gather family members in the bathtub; pull a mattress over your head; and hope for the best.

At 5:05 A.M. Monday the hurricane came ashore, with sheets of rain and sustained winds of 145 miles per hour, gusting to 175. Ocean water, pushed by the heavy winds, surged over the land, in a tide as much as fifteen feet deep. At the hurricane center's headquarters south of Miami, rooftop instruments blew away. Later, scientists tracked the storm's path using readings from instruments elsewhere in the state, including scores of measurements contributed by amateur meteorologists who answered the hurricane center's broadcast appeals for help. "Remarkably, more than one hundred quantitative observations were received," the center said later, in its official account of the storm. "Many of the reports came from observers who vigilantly took readings through frightening conditions including, in several instances, the moment when their instruments and even their homes were destroyed."[3]

Andrew crossed the state in four hours, weakening a bit. But the storm gained strength again as it moved across the warm water of the Gulf of Mexico. By now, a hurricane watch had been posted for the entire coast of Louisiana, from Mobile, Alabama, to Sabine Pass, Texas. In low-lying places south of New Orleans, families escaped on buses driven by members of the Louisiana National Guard.

"It was as if a bomb had been dropped on Louisiana," Colonel William J. Croft, an official in Louisiana's Office of Emergency Preparedness, recalled later. "I know it's dropped, now I'm waiting for it to explode. My mind was trying to visualize eighteen to twenty feet of water in downtown New Orleans and what would we do with the people down there. . . . We were planning for the worst case."

The hurricane looped northwest across the Gulf to the Atchafalaya River basin southwest of New Orleans, where it turned north and moved inland before dawn on Wednesday, August 26. By then it had weakened to a Category 3 storm, but it was still powerful enough to devastate the fragile barrier islands along the coast there. So much sand was removed from their ocean beaches that they virtually rolled over. "These islands have such low relief that, with the storm surge and the waves, sand was just lifted up and carried over and deposited on the back side," S. Jeffess Williams, a coastal scientist at the United States Geological Survey, said later. "The sand was not removed from the front and put on a flood tidal platform on the island. It was put behind the island. This is an extreme form of rollover."

But by then the storm's fiercest energy was spent. Andrew moved northeast and finally ran out of steam two days later, merging with frontal systems over the Middle Atlantic states.

The *Miami Herald* called Hurricane Andrew "the big one," and by some measures it was. Measured by low pressure at the eye, a rough guide to storm intensity, Andrew was the third strongest hurricane ever recorded in the United States. (The two strongest were a 1935 storm that wrecked the Florida Keys and Camille, the only Category 5 storm to make landfall in this country in this century, which struck the coast of Mississippi in 1969.) Andrew cost fifty-two lives and caused almost $30 billion in damage to homes and businesses. Tens of thousands of families were left homeless.

But despite the headlines, Andrew was not the storm of meteorologists' nightmares. It is a worst-case storm only because people who survived it now think—mistakenly—that they survived the worst that Nature can do. In some ways, it might even be described as a best-case hurricane for Florida and Louisiana.

The hurricane made landfall not at Miami, where glass-walled condos rise shoulder-to-shoulder on the beach, but rather about twenty miles to the south, in Biscayne National Park, where the muddy shoreline, lined with mangroves, has never been attractive to developers. In most of Miami, the wind was barely hurricane strength, and in northern Dade County the biggest problem was downed power lines. To be sure, the storm destroyed hundreds of homes in the neighborhood of Homestead Air Force Base, and nearby nursery operations lost their stock. Residents of the area suffered mightily. But they were relatively few. With the exception of the base, already listed by the Pentagon for closing, no major infrastructure lay in the storm's path as it chewed its way through the Everglades. Andrew crossed into the Gulf of Mexico not at populous Tampa or Fort Myers but much further south, along a stretch of coast that was, by Florida standards, sparsely populated.

As the hurricane moved out over the warm Gulf, gathered strength, and headed north, disaster officials watched with bated breath. Again their luck held. When Andrew made landfall in Louisiana, it came ashore on the state's muddy chenier coast, well to the west of New Orleans. Here too development was relatively sparse—the washed-over barrier islands were actually uninhabited.

"Had the track been twenty miles to the north—and meteorologically we cannot even tell you the difference that would have created that track—downtown Miami would have looked like Homestead," Robert C. Sheets, then head of the National Hurricane Center, told the seventeenth annual National Hurricane Conference in 1995. "And if it follows our nightmare track into New Orleans, it drives water into Lake Ponchartrain and empties it into New Orleans. The city would be under eighteen to twenty feet of water. Had Andrew been on that northern track in New Orleans, Florida would have been an afterthought." Damages would have mounted to more than $100 billion. (Disaster officials make the same point about Hurricane Hugo, which struck South Carolina in 1989. Though it was ferocious, it too made landfall in a relatively unpopulated area, at Francis Marion State Forest.)

Hurricane Andrew should be remembered as a narrow escape, a warning about what can happen when people overdevelop the coast. But that is not how people remember it. "Some of the citizens who lived through

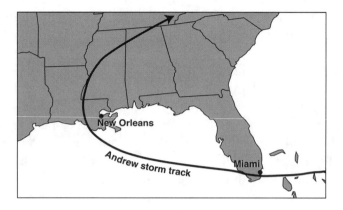

The track of Hurricane Andrew missed heavily developed areas. A shift of even 20 miles north would have made the disaster far worse. *(National Hurricane Center)*

it with no effects will say, 'Here comes another Category 4 hurricane. I can live through it,' " Croft said. "It's complacency, it's a false sense of security. The next storm is going to tell us different stories."

When the deadly hurricane struck Galveston in 1900, the only advance information the Weather Service had about the storm came from haphazard ship reports. As late as 1938, when a powerful hurricane struck Long Island and New England, more than six hundred people died, many because they did not realize a powerful storm was bearing down on them. Meteorologists' ability to forecast the movements of hurricanes has improved tremendously since then, but population growth along the coast is more than keeping pace. Almost everywhere on the East and Gulf coasts, by the time the meteorologists can speak with certainty about where a hurricane will strike, it is far too late to order an evacuation. There are simply too many people to move. This situation is only getting worse. For example, Florida's coast was home to about five million people in the 1960s. By 2010, disaster officials estimate, sixteen million people or more will live there.

The problem is especially acute on barrier islands, where escape routes may be limited to one or two low-lying causeways whose approaches flood hours before a storm actually hits. Seven bridges connect the barrier island towns of Dade County to the Florida mainland, but all of them are approached by low-lying causeways and six are drawbridges, vulnerable to mechanical problems in storms. The Florida Keys, a 140-mile string of islands where development has been intense, are particularly dangerous. Their average elevation is three feet and 90 percent of their land is less than five feet above sea level. There are seventeen Red Cross shelters there, but most of them would be flooded in a Category 3 storm.

Other islands, like Ocracoke in North Carolina, or Fire Island on the south shore of Long Island, can only be reached by ferry—and high winds may force ferry companies to cease operations a day or more before a storm hits.

Evacuation can be difficult to achieve even in places where road access is relatively good. Meteorologists can only give eight to twelve hours of reasonably sure warning that a storm will strike. For much of the coast, that is not enough time. Public safety officials in Florida estimate that it would take thirty hours or more to evacuate Fort Myers, the Florida Keys, or the Palm Beach area.[4] After Hurricane Andrew, Sheets said officials in Louisiana, particularly in New Orleans, should have forced people to leave the city for shelter elsewhere. But city officials angrily rejected this criticism, saying timely evacuation is impossible in New Orleans. In part because many residents do not have cars, officials said, they would have to have ordered an evacuation seventy hours or more before the hurricane hit—when it was still a tropical storm in the Caribbean. Sheets himself has conceded that it would probably take more than eighty hours to evacuate the southeast Florida coast.[5]

The situation is further complicated because evacuation orders must be issued when people are awake to hear them, and they should be carried out in daylight (preferably at low tide) and before the onset of the heavy rains that signal the storm's approach. Rarely do all of these factors line up. When Hurricane Opal was approaching the Florida panhandle on October 3, 1995, its winds were barely hurricane strength. But they strengthened dramatically overnight. At sunrise on October 4, the storm winds had reached 150 miles per hour.[6] By then, bridges were closed and Interstate 10, the major panhandle evacuation route, was jammed with cars. Fortunately, the wind dropped as the storm moved on shore, but even so Opal was still one of the most destructive in Florida history.

The question of how soon, and how forcibly to order an evacuation has become one of the most intractable public safety problems on the East and Gulf coasts. Most officials are unwilling to issue evacuation orders, which are costly to states and localities and to private citizens who must leave work, lose business, and seek accommodations elsewhere. According to the National Oceanographic and Atmospheric Administration, the average warning generates $63 million in evacuation and preparation costs. Others put the cost at $1 million per mile of coast.[7] If the storm shifts and bypasses the evacuated area, many people will be furious at having borne the effort and expense of evacuation for nothing—and they will be far less willing to heed the next evacuation order.

But if officials wait until they are pretty sure where a storm will strike— usually about eight to twelve hours before it makes landfall—it is far too late for people to escape. "Sometimes you overreact," Sheets said in an interview. "Sometimes you underreact. We hope that people err on the high side, but there's a genuine concern on the part of government—will the people heed the evacuation order?"

Though the science of predicting hurricane movements is improving, it remains very much an art. When tropical storm Gordon formed in the Gulf of Mexico in 1994, computer models initially predicted several possible tracks. As it moved across the Gulf, the computers narrowed the prediction to a Florida landfall in seventy-two hours, with winds of more than one hundred miles per hour. By then, heavy rains were falling on south Florida and the storm was moving at almost twenty miles per hour. Suddenly, it stopped in its tracks and turned around. ("I told the reconnaissance people to look for skid marks on the ocean," Sheets recalled at the hurricane conference.) The storm crossed into the Atlantic and then made several more sharp course corrections before petering out off New York. "It was at this stage," he said, "that I turned in my resignation."

By the turn of the century, officials at the hurricane center say, new computer models, radar equipment, and aircraft will add enough precision to improve hurricane forecasts by 20 percent or more—that is, the forecasters will be able to narrow the warning areas by 20 percent. But they concede that such improvement, though substantial, will not be enough to solve the problem. And these improvements will occur only if Congress appropriates the money to fly the new airplanes and run the new computers.

Once an evacuation has been ordered, it must be carried out, and that is no easy matter. Most coastal towns are clogged with traffic in summer. An evacuation order only makes highway congestion worse. The governor of South Carolina solved this problem when Hurricane Hugo was bearing down on Charleston by ordering state police to reverse traffic flow on part of Interstate 26, turning the highway into a gigantic one-way street headed inland. Facing similar traffic problems, public safety officials elsewhere on the coast are planning to do the same thing.

Unfortunately, by the time road builders began to realize such steps might one day be necessary, much of the coastal highway system was already in place. Had designers thought about it in time, they might have designed on- and off-ramps, interchanges, and other features so that they could easily change direction in case of emergency. Some states plan to modify their highways, to retrofit them with as many of these features as they can. For example, Croft says Louisiana plans to build highway crossovers for use in such a situation.

Meanwhile, though, other jurisdictions are planning how to make do with what they have. In coastal New Jersey, where villages such as Avalon, population eighteen hundred, swell to cities of forty thousand-plus in the high season, officials have long known their highway system would be inadequate in the event of an emergency evacuation; roads there are already clogged with traffic on ordinary summer days. And if people delay their departure—as Jerseyites, few of whom have experienced a hurricane at firsthand, are likely to do—roads will be overwhelmed with people trying to make a last-minute escape. In 1993, emergency management officials and state police conducted a study of evacuation. They found it would take thirty-six hours, far too long, to evacuate Cape May County, the beach resort area at the state's southern tip. But they could cut that time in half by turning the Garden State Parkway, a major north-south highway, into a one-way road out.

Since then, the state has made plans to put this emergency measure into effect. Officials have stockpiled highway barrels and sandbags in strategic locations; painted hundreds of one-way signs and stored them at the ready; identified tow trucks and front-end loaders to be drafted to keep roads clear; stockpiled variable message signs; and purchased other supplies, everything from mobile phones to duct tape. The one-way plan would go into effect as soon as the state's governor declared a state of emergency because of a storm. But turning a major highway into a one-way escape route is a high-stakes gamble. The effort requires scores of police officers and platoons of tow-truck operators, because even a single broken-down car could back up traffic for miles.

In some areas, though, even an ad hoc redesign of the highway system would not be enough to ensure a timely evacuation. Officials in Florida are wondering whether it would be better, overall, not to evacuate Dade County when flooding is predicted to be less than a foot or two. Florida has severe evacuation problems, especially if a storm crosses the state. People escaping both coasts would end up on highways converging at Orlando, a traffic bottleneck of potentially enormous proportions. And the highway is about the worst place to be in a hurricane.

All of these problems are leading some preparedness officials to a conclusion that would have been unthinkable even a few years ago: maybe people should be encouraged to ride out the storm where they are. Up to now, emergency management officials have derided this approach as foolhardy, citing as their favorite example the dozens of people in the Richelieu Apartments in the coastal town of Pass Christian, Mississippi, who held a hurricane party as Camille approached the Mississippi coast in 1969. Only one survived. But few officials cite this example any more. They recognize that there are too many people on the coast who will not be able to escape even if they want to.

As a result, many coastal towns are establishing what they call "refuges of last resort," shelters people could seek out if the storm was upon them and they had nowhere else to go. These places would not be comfortable. They would not even offer the amenities of a standard Red Cross shelter. But they would at least offer some protection from the storm. Schools, firehouses, libraries, or other public buildings are first choice for refuges of last resort. (Among other things, the threat of liability litigation is greatly reduced, and municipal officials can order them to be equipped for use.) But movie theaters, churches, hotels, or clubhouses might also suffice.

Planners consider many factors in deciding whether a particular building could serve as a refuge of last resort. It should be built to code and be in good repair. It should be near major evacuation routes and away from open bodies of water or even drainage ditches. It should be on high ground so that even if it floods it is not subject to wave action. (In many barrier islands, however, there is no high ground.) There should be no towers, trees, or other structures nearby to fall on it. And it should be equipped in advance with generators or at least some form of radio or wireless phone communication. Habitability standards are low: nine square feet per person—about enough room for an adult to lie down—and a few portable toilets. People would have to bring their own supplies, maybe even their own water, and they would be uncomfortable until the storm passed. In the north, where hurricanes tend to move swiftly inland, that might not be more than a few hours. Further south, though, hurricanes move more slowly, and people might have to stay in their refuge for a day or more.

Two of the first jurisdictions to establish refuges of last resort were the Florida Keys and the city of Galveston, both of which have experienced storms strong enough to destroy them. During Hurricane Andrew, people on the Keys who did not pass a highway checkpoint by a time determined by traffic conditions were directed instead to designated refuges.[8] Other localities are taking action, too. For example, Sanibel, Forida, a heavily developed Gulf Coast barrier island linked to the mainland by only one bridge, now requires all buildings open to the public and containing more than three thousand square feet to be equipped as refuges of last resort.

Unfortunately, weather is rarely a factor when developers and municipal planners consider when they sit down to discuss building on the beach. Though meteorologists can produce years of records showing that hurricanes, northeasters, El Niño storms, and the like occur regularly on the coast, builders and planners still make decisions as if they were unpredictable acts of God rather than business as usual. This way of thinking became all the more entrenched on the East Coast in recent decades, when a surge in building occurred in a time of unusually low hurricane activity. According to Sheets, if hurricane frequency returned to the pattern of the 1940s, 1950s, and 1960s, multibillion-dollar disasters would be yearly events.[9]

Nowadays there is a lot of talk that global warming will intensify storm activity. That remains to be seen. But from a meteorological point of view it makes sense: storms such as hurricanes draw their energy from the heat in the ocean; if the oceans warm, there will be more energy to power storms. In any event, the era of low storm frequency seen for the last few decades cannot be expected to continue indefinitely. One scientist who studies patterns of hurricane frequency, William M. Gray of Colorado State University, has tied them to the droughts that come and go in the Sahel, the desert area in West Africa. This region is the nursery for weather disturbances that eventually coalesce off the African coast and move across the Atlantic to become hurricanes. When the Sahel is dry, few major hurricanes hit the East Coast of the United States. From 1966 to 1990, a dry period in Africa, only three hurricanes of greater than Category 3 intensity struck the United States. In the previous twenty-five years, a wet period, there were fifteen.

Weather factors that reduce storm frequency in one place may increase it somewhere else. For example, in El Niño winters, when currents of warm water flow in unusual patterns in the Pacific, related changes in the jet stream can interfere with hurricane formation in the Atlantic, reducing the odds of a catastrophic storm hitting the East Coast. But El Niños are disastrous on the West Coast, where they can produce a winter full of heavy rain and pounding surf to weaken and erode ocean-side bluffs.

In fact, though they get the most press, hurricanes are not the biggest weather threat to the beach, if only because they are relatively infrequent, even in a bad hurricane year. That is not the case with the northeasters that pound the East Coast, the cold fronts or "northers" that move across the Gulf Coast, or the winter storms that batter the Pacific. People often describe these storms as "freak," but they are far from it. In a typical year,

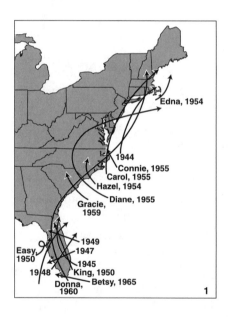 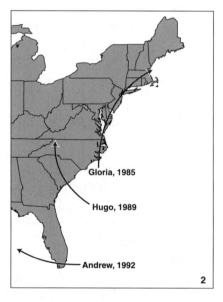

Hurricane patterns on the East Coast of the United States seem to be related to whether the weather is wet or dry in West Africa, the nursery of hurricanes. The last wet period, from 1944 to 1960, produced many strong hurricanes. From 1960 to 1992, weather in Africa was dry, and strong hurricanes were few. *(Florida Coastal Management Program/Jen Christiansen)*

at least thirty northeasters approach the Atlantic coast of the United States with enough force that their waves erode ocean beaches. On average, virtually every stretch of U.S. coast is struck by a severe storm every few years. And unlike hurricanes, which pass so quickly they usually do comparatively little damage to the beach itself, big winter storms can linger for days, marking the coast for years or even decades.

One of the worst, and still a benchmark against which other storms are measured, was "the great Ash Wednesday storm," a giant northeaster that raged along one thousand miles of the Atlantic coast from March 5 to March 8, 1962. The storm struck at the perigean spring tide, when the tides rise highest, and howled through five high tide cycles (some people still refer to it as the "five high storm"), with waves as much as thirty feet high. The Ash Wednesday northeaster destroyed the famous Steel Pier of Atlantic City, New Jersey, and the boardwalk at Ocean City, Maryland, and dumped thousands of beach houses into the surf.

On barrier island after barrier island, beaches were eroded back to the dune line, and roads and house lots were buried in tons of overwashed sediment. Inlets were choked with sand and debris. Before it wore itself out, the storm had killed thirty-two people, left hundreds injured, and caused more than $500 million in damages—in 1962 dollars.[10] The economic blow such a storm would strike today, with much higher property values and much more intense development, is almost too terrible to contemplate. Certainly, damage would be in the billions.[11] (The so-called Storm of the Century, a northeaster that struck much of the East Coast in March 1993, caused billions of dollars in damage, though it was much weaker than the Ash Wednesday storm.)

What struck coastal scientists after the Ash Wednesday storm were the glaring differences in the ways developed and undeveloped islands responded to the storm. The storm had extensively rearranged the Atlantic coast, but undeveloped beaches shook off its effects. That was not the case in developed areas. More than anything before it, the storm convinced many coastal scientists that development at the coast, particularly development on barrier islands, was hazardous for people and bad for beaches, too.

"It became clear following the March 1962 storm that understanding the natural dynamics of beaches and barrier islands is important not only from an academic standpoint, but it is also the key to recognizing and estimating both the short-term and long-term hazards of living on them," scientists wrote.[12] The storm turned out to be a major catalyst for coastal research. Unfortunately, the lessons of this research have yet to be put into practice.

One factor is the widespread confusion over how to compare the severity of one northeaster against that of another. The question is far from academic; in fact, it is crucial in the debate over the usefulness and feasibility of beach nourishment as a technique for coastal protection. When a storm washes away a replenished beach, it is important to know whether the beach vanished under the onslaught of an unusually severe storm or the kind of bad weather that can be expected to occur every few years.

Since 1971, hurricanes have been rated according to a scale devised by Herbert S. Saffir, a Miami engineer, and Robert Simpson, then director of the National Hurricane Center. This scale compares hurricanes by wind speed, barometric pressure, and the storm surges they produce. Category 1 hurricanes, the weakest, have winds of at least 74 miles per hour and barometric pressure greater than 28.91 inches, and typically produce storm surges of four to five feet. Damage from these hurricanes is minimal, and it is usually limited to trees and bushes, outdoor furniture, or mobile homes. Category 5 storms, the fiercest, have winds in excess of 155 miles per hour and barometric pressures less than 27.17 inches. Category 5 storms are typically accompanied by storm surges of eighteen feet or more, and the damage they cause is catastrophic.

Scientists are not completely satisfied with the Saffir-Simpson scale, especially when it comes to storm surge, which can vary dramatically according to how fast the storm is moving and the topography of the ocean bottom and the shoreline where it makes landfall. According to Sheets, the same Category 4 storm would produce a surge of ten feet in Miami Beach, twenty-five to thirty feet on the Florida panhandle, and eighteen to twenty feet in New Orleans.

Assessing nonhurricane storms such as northeasters is even more difficult. Usually, scientists compare them by estimating the odds that a storm that size will occur in any given year. Coastal scientists call this figure the storm's "return interval." For example, a storm with a 5 percent, or one in twenty chance of occurring in a given year is called a twenty-year storm. A storm with a 1 percent chance of occurring has a one-hundred-year return—it is truly a "storm of the century."

These calculations can be difficult to make. Recently, Robert Dolan, a marine geologist, and Robert E. Davis, a climatologist, both at the University of Virginia, devised a scale for use in comparing northeasters. Their scale divides storms into five classes according to factors such as the amount of beach and dune erosion they cause, their duration, and their peak wave height. A Class 1 storm would last about eight hours and have peak waves of just under seven feet, on average. It would produce minor beach erosion without any dune erosion or overwash. This kind of storm has a return interval of perhaps two weeks to a month, in the winter. A Class 3 storm would produce erosion across the beach and significant dune erosion. Average peak wave height would be almost eleven feet, with the storm lasting perhaps thirty-four hours. The most severe storm, a Class 5, would last for four days or more, with wave height peaking at twenty-

Storm Class (Northeasters)

	Class 1 (weak)	Class 2 (moderate)	Class 3 (significant)	Class 4 (severe)	Class 5 (extreme)
Beach erosion	Minor	Modest	Across beach	Severe	Extreme
Dune erosion	None	Minor	Significant	Erosion and recession	Destruction
Overwash	No	No	No	Severe on low-profile beaches	Massive, in sheets and channels
Property damage	No	Modest	Local	Community wide	Region wide
Average peak wave height (ft.)	6.6	8.2	10.8	16.4	23
Average duration (hrs.)	8	18	34	63	96
Relative frequency (odds of occurring per year	49.7%	25.2%	22.1%	2.4%	0.1%

Scientists at the University of Virginia devised this scale for assessing the relative power of northeasters. *(Robert Dolan)*

Saffir-Simpson Scale of Hurricane Intensity

Category	Wind	Barometric Pressure	Storm Surge	Potential Damage
1	74 to 95 m.p.h.	More than 28.91 inches	4 to 5 feet	Minimal: Damage to trees, shrubbery, unanchored mobile homes.
2	96 to 110	28.50 to 28.91	6 to 8	Moderate: Some trees blown down; major damage to exposed mobile homes: some damage to roofs.
3	111 to 130	27.91 to 28.47	9 to 12	Extensive: Trees stripped of foliage; large trees blown down; mobile homes destroyed; some structural damage to small buildings.
4	131 to 155	27.17 to 27.88	13 to 18	Extreme: All signs blown down; extensive damage to windows, doors, and roofs; flooding inland as far as six miles; major damage to lower floors of structures near shore.
5	Greater than 155	Less than 27.17	Greater than 18	Catastrophic: Severe damage to windows, doors, and roofs; small buildings overturned and blown away; major damage to structures less than 15 feet above sea level within 500 yards of shore.

three feet. Such a storm would produce extreme erosion, would destroy dunes, and would send massive amounts of sand washing inland. This is the one-hundred-year storm.[13]

Their system has yet to come into wide use. Some coastal scientists praise it for factoring a storm's duration into calculations of its punch, but others say doing so can produce anomalous effects. For example, on the Dolan-Davis scale, the Ash Wednesday storm is only the fifth most intense to strike the East Coast. Though it was extremely violent, it came and went relatively quickly. Still, there is no doubt that accurate comparisons of northeasters would be a boon to coastal officials and emergency planners.

Northeasters are a regular weather event only on the East Coast, but bad weather stalks the beaches of the West Coast, too, particularly in El Niño years. And even though El Niño events have left spectacular destruction in recent years, records and accounts from the early nineteenth century show that California weather then was, if anything, worse.

In his book *Two Years Before the Mast*, Richard Henry Dana describes the terrible storms of those years, when ships plying California's coastal waters were warned to be ready to weigh anchor and head for deep water almost at any moment, so fierce and unpredictable was the weather. Sailors described the winds as worse than those at Cape Horn, producing waves fifty to sixty feet high. Nothing like this kind of weather has occurred in the twentieth century, not even in the winters of 1982–3 or in the 1990s when disturbances in Pacific currents produced El Niño winters of harsh storms.[14]

Besides threatening shipping, these storms steadily ate away at the coastline, and they are one reason why ancient maps of settlements such as San Diego record many streets that no longer exist. Scientists who have studied these old weather patterns say there is no way of knowing if they were a onetime anomaly or if the storms might recur, with far greater economic cost today than 150 years ago.[15]

The death toll from hurricanes striking the United States has plummeted since the Galveston storm of 1900. But the toll in property damages has exploded. The National Hazards Research Center, a federal agency, estimates that between 1975 and 1994 the United States spent an average of $250 million a week on natural disasters, many of them on the coast. Hurricanes alone cause billions in damages.

According to the National Oceanic and Atmospheric Administration, the ten costliest hurricanes (not adjusted for inflation) were:

Andrew (1992): $24 billion
Hugo (1989): $7 billion
Frederic (1979): $2.3 billion
Agnes (1972): $2.1 billion
Alicia (1983): $2 billion
Iniki (1991): $1.8 billion
Bob (1991): $1.5 billion
Juan (1985): $1.5 billion
Camille (1969): $1.42 billion
Betsy (1965): $1.42 billion[16]

These hurricanes were destructive largely because too much building has happened in unsafe coastal sites. But the situation could be improved substantially if local authorities adopted and enforced adequate construction standards. Requiring a few relatively simple and inexpensive construction steps can make a big difference in the cost natural disasters exact. Better-built buildings survive the storm and, by surviving, they offer little or no debris for storm winds to fling around at other buildings, a major cause of damage in hurricanes. After Hurricane Andrew, a survey by the Federal Emergency Management Agency found extensive damage "caused and further promoted by airborne debris. The debris consisted largely of failed roofing materials, but also included components of metal-clad buildings, various accessory structures, and miscellaneous sources such as fences."[17] Anything that could become airborne became an instrument of destruction.

Andrew also revealed shoddy construction and inadequate building code enforcement throughout South Florida. "Substandard workmanship was noted at many locations," the FEMA report said. "Clearly, not all tradespeople were well qualified in the construction of building structural systems, structural components, and connections necessary to resist design wind loads. Where high-quality workmanship was observed, the performance of buildings was significantly improved." But the report was scathing in its assessment of building inspection in many areas. "Inspection was inadequate to address the workmanship problems observed," it said, particularly in large developments in which structures have similar designs and inspection is a repetitive process.[18]

In some jurisdictions, solving this problem will require rewriting building codes. This is because the goal of most building codes is the construction of houses that will stand up to gravity—that is, houses in which the beams hold up the ceiling and the joists hold up the roof. With storms, though, the problem is wind not weight. Even codes that contain wind standards often set them at 120 miles per hour. But in Hurricane Andrew there were places in South Florida where sustained winds were 140 miles per hour, with gusts of 170 miles per hour. "Sometimes you have to exceed codes unless the building or its contents are expendable," Sheets said.

Hurricane experts and civil engineers estimate that a house can be made hurricane-resistant (no one who knows about coastal storms would call any building "hurricane-proof") with relatively modest steps like using reinforced windows, heavy-duty roofing nails, or metal braces called "hurricane clips" to attach wall joists to roof beams, and the like. Even adding a nail or two per roofing shingle can cut down damage dramatically. Architects and engineers say improvements like these vastly increase the ability of a house to survive a hurricane, yet they can be added at an additional cost of only about 2 to 10 percent of the value of the structure. Perhaps more important to surviving a hurricane, though, is designing the house without some of the features—like overhanging roofs—that make it easy for wind to tear a building apart. Other features that can cause problems are large rooms with cathedral ceilings, which offer little support for the structure of the house itself, large expanses of glass, which mean weaker walls, and nonfunctioning shutters.

At a conference on hurricane preparedness after Andrew, Sheets described a helicopter tour he made of the swath of South Florida struck by the storm. On street after street, he said, he could see places where storm winds had gotten under eaves and literally peeled the roof away. One development was particularly hard-hit, he said, because its houses had been constructed with plenty of eaves and overhangs. He showed before and after photographs of one heavily damaged house, originally designed with a sweeping overhang, remarking, "If you wanted to design your house to fly, this is the kind of roof you put on it."

He went on, "We had gone for twenty years without a major hurricane in South Florida. We started to get northern-style homes. It's almost as if the architects decided, 'How can I catch the wind?' It's like they did their best to design a building that will catch the wind." He showed photographs of another house, half of which had been torn away. "It was well constructed," he said. "It wouldn't be standing otherwise. It was good construction but very poor design."

As part of a nationwide campaign to encourage the building of disaster-resistant housing, FEMA and a number of private companies—makers of windows, home repair chains, and other concerns—have paid for the construction of such a house in the small town of Southern Shores, North Carolina, on the northern stretch of the Outer Banks. Interior walls of the house are of Plexiglas so builders and their prospective customers can see how it is done. Among other things, the exterior of the house is designed without overhangs that catch the wind, and the windows are constructed of two panes of glass with a sheet of reinforcing clear plastic in between. The insurance industry estimates that $5 billion in claims after Hurricane Andrew covered water damage resulting from broken windows. Intact windows also reduce by half the chances that hurricane winds will get inside the house and explode it, blowing the roof away.[19]

The project also aims to find relatively inexpensive ways homeowners can retrofit houses built without these features. For example, the project is supporting the design of equipment to spray foam sealant under roofs, to strengthen them. Eventually, participants hope to be able to help people strengthen joints, perhaps through small holes in interior walls that might be covered by molding.

Still, even a big storm does not necessarily make people storm-conscious. At the hurricane conference, Sheets also spoke of a trip he made to South Carolina after Hugo, where he saw house after house being rebuilt without hurricane clips. Occasionally, he said, he would see people apparently attempting to reduce their future losses by rebuilding with cinder block and steel beams, "but no engineering—just trial and error."

Finally, there is the question of mobile homes, a popular housing choice in many coastal towns in the South, particularly among retirees on fixed incomes. It is possible to make them safer by installing new footings, hurricane clips, cables, and the like, as a FEMA report recommended in 1985, but this retrofitting is often beyond the economic reach of low-income homeowners. And even with this reinforcement, mobile homes are still highly dangerous in a storm. Many disaster officials think they should be banned entirely from the coast. In its report after Andrew, FEMA recommended renewed state and federal studies of design standards for manufactured homes but added, "The issue of providing safe, affordable housing in high-wind areas needs to be further examined."[20]

Pilkey, his associates at the Duke University Center for the Study of the Developed Shoreline, and coastal scientists they recruited from around the country have produced a series of books on "How to Live with" various reaches of the Pacific, Gulf, Atlantic, and Great Lakes coasts. State by state, the books describe coastal processes in general, processes unique to or prevalent in particular areas, and the storm history of that stretch of coast. Then each book offers detailed mile-by-mile analyses, providing prospective buyers with information they need to avoid settling in high-hazard zones and advising property owners already there of steps they could take to improve their prospects in the event of a storm.

Often, these steps amount to little more than undoing damage developers or subsequent owners have done in order to take fullest economic advantage of a site. For example, the books advise replanting trees and shrubs and rebuilding dunes. Though these natural buffers are ideal storm protection, builders often remove them to improve a structure's view of the ocean—and raise its selling price. But as Pilkey puts it, "If you can see the ocean, the ocean can see you." Modest dune-building and replanting offers little protection against a big storm, but it can help in smaller storms. If possible, plantings should be of native vegetation, more likely to thrive and survive in a storm.

One surprisingly effective tactic for strengthening barrier islands is to encourage the formation and growth of marshes on their inland sides. Once a marsh is established, even in a small area, its grasses will start to trap sediment that might otherwise end up in the middle of the sound or lagoon, and the more sediment is trapped, the more the marsh will expand. The result can be effective shore protection without any of the habitat or nutrient loss that occurs when people seal off the marsh shoreline with bulkheads.

Other mitigation steps are more effective but would be much more disruptive and expensive. For example, rerouting barrier island streets so that they curve, rather than running straight from ocean to bay, can deprive the ocean of overwash channels that would otherwise fill with water in a storm and flood the island behind the dune line.

Overall, communities need long-term plans to encourage building away from the beach and to block people from rebuilding in hazard areas. In particular, they should plan to relocate or abandon infrastructure—roads, power lines, and the like—should they be damaged in storms.

Finally, they should identify hazard areas, particularly inlet hazard areas, so that buyers know what they are getting into when they purchase coastal property. For example, there are many barrier beach towns where houses have been built where inlets once flowed. If local officials ordered the inlet to be closed artificially, often the case nowadays, it probably had no time to build up the ebb and flood tide deltas of sand that make an island stronger. In any case, as long as the natural offshore or bayside factors remain in play, an inlet may form there again in the next storm.

From time to time, scientists and coastal planners have proposed mapping the shoreline to identify high erosion areas. So far, though, real estate and other commercial interests have been able to block the idea. They have the most to lose.

CHAPTER EIGHT

Clues

Let Nature be your teacher.
—William Wordsworth, "The Tables Turned"

Before dawn on a gray October morning in 1990, two dozen scientists crowded into a cold trailer parked in the dunes of Duck, North Carolina, sipping coffee from paper cups and hoping for heavy weather. They had come to this remote spot from all over the country to match wits, muscle, and technology with the ocean. Their challenge was a simple question: Why are the nation's beaches eroding?

They already knew the short answer. Rising seas were moving in on the coast, and people were interfering with normal sand movement, preventing the beach from defending itself. But there was much they did not know about *how* erosion occurs, because there was much they did not know about exactly what happens when air, land, and water meet at the beach. To find out, they had gathered at a small research establishment run by the Army Corps of Engineers. Formally, this establishment is called the U.S. Army Corps of Engineers Waterways Experiment Station Field Research Facility. Scientists know it as the Duck pier.

Though fall is usually warm on the Outer Banks, the air was chill in the morning darkness when the oceanographers, geologists, computer scientists, and engineers convened. They had constructed and installed dozens of instrument arrays in the surf and farther offshore and, one by one, they described how each was faring. A hurricane was making its way up the coast and they wanted to be ready to measure its effects.

Days of heavy surf, a harbinger of the approaching storm, had already altered the shape of the beach, and the changes were displayed in graphs, maps, diagrams, and computer printouts taped to the trailer's brown panel walls. Detailed topographical diagrams, covering the shore from the dunes through the surfzone, showed that the beach was shifting some of its sand into a nearshore bar that would break the waves before they hit the beach. Irregular at first, the bar now seemed to be organizing itself to meet the storm.

"The bar is going linear," William A. Birkemeier, the engineer who runs the facility, told the group. "The surf is up. The instruments are being pounded."

As chief of the Field Research Facility, it was up to him to make sure the scientists had what they needed to collect their data. One by one, they told him about their problems and successes of the day before and outlined their goals for that day and the next. As each spoke, Birkemeier made notes of what had to be fixed, found, moved, removed, or installed.

Ed Thornton, a professor at the Naval Postgraduate School in Monterey, California, and an expert on nearshore currents, reported on instruments he had mounted on in the fixed array of pipes and cables running into the surf. Water-pressure sensors to measure the waves were okay, he said, but the current meters, originally arrayed in parallel, were a problem. They were turning toward each other. Thornton needed a diver to go into the surf to adjust them.

The others in the trailer raised additional problems—dud fuses, or wave gauges that had suddenly fallen silent. Bob Guza, from the Scripps Oceanographic Institution, said he would have the Scripps divers reorient some sensors he had mounted in an offshore array. "They are producing anomalous data," he said. Guza, who studies underwater acoustics for the navy, among other things, is one of the nation's leading coastal scientists, but he hardly looked the part this morning. He was dressed in shorts and a white T-shirt with elaborate black lettering running around from front to back—the formula for calculating the sticking power of a barnacle. His calloused feet were bare.

As each problem was raised, there was a brief discussion and then a decision on how and when it would be solved. Birkemeier concluded the meeting by announcing that there would be a resupply mission to Norfolk, Virginia, the nearest big city, seventy miles to the north. Everyone had something to put on the shopping list—rope, cable, electrical equipment, computer supplies. Shortly after 8 A.M., the scientists discarded their empty coffee cups and headed out into the morning, ready for work.

When the pier began operating in 1978, the village of Duck was little more than a bend in the narrow road that runs along the Outer Banks north of Kitty Hawk. Its sand dunes were covered with wax myrtle and scrub oak and its sound-side marshes were thick with the waterfowl that gave the village its name. The birds are much fewer nowadays and the sand dunes have sprouted street after street of high-rent vacation houses, but the 176 acres of corps land, running across the barrier island from ocean to sound, are much as they were.

Twelve engineers, scientists, and technicians work full-time here, maintaining an impressive array of instruments, including many they designed and made themselves. These gauges, sensors, and meters are mounted on buildings, hung from a 140-foot metal tower and tethered to a pier that runs more than a quarter mile beyond the surf, where the water is 35 feet deep. Day after day, in fair weather or foul, the instruments read the winds, waves, and currents, and record their changes. Still other instruments, towed into the surf, map changes in the shape of the beach.

Over the years, the Duck pier has compiled an exhaustive record of the natural history of this stretch of beach. So this is where coastal scientists go when they want to test new theories or try out new instruments. Many of them say the Duck pier is the best place in the country, maybe the world, for field research on coastal geology. Their results—previously unknown currents detected, or sandbar patterns observed for the first time—regularly surprise even themselves. Each discovery prompts new rounds of research. Most of the discoveries come when scientists gather, as they had this October, to pool their resources against the ocean.

They need each other to maintain the necessary array of instruments, a constant worry in coastal research. It is difficult to design sensors and gauges to be at once accurate, quick to respond to changing conditions, and sturdy enough to survive in the surf. They are also vulnerable to vandalism and damage from fishing boats, which is another reason why scientists like to work at the pier—the corps can limit public access to the beach and nearby water.

That evening found Birkemeier cruising the beach in an open jeep, looking for an instrument canister that had broken loose from one of the arrays. The missing canister held about $20,000 worth of instruments, not to mention the records of data the scientists did not want to lose, and Birkemeier hoped he would find it—intact—washed up on the high tide line. About half a mile north of the pier, he spotted the dark, two-foot cylinder, lying in a clump of sea oats at the edge of the dune. The canister was still in one piece but it felt suspiciously heavy, suggesting it was waterlogged. "There's no sloshing so maybe it's okay," Birkemeier said hopefully. In growing darkness, he cast an appraising eye at the waves, which had strengthened noticeably since morning, just as the scientists had hoped.

"Looks like it's going to rain," he said happily. "Better batten down the hatches."

Beaches are like icebergs, coastal geologists say. Ninety percent of the action is under water. But, as with icebergs, studying the underwater portion of the beach is dangerous and difficult. As a result, much remains unknown about the currents that run along the shore, and the movement of sand within them. The surfzone is still a realm of mystery. In many ways, the basic work of learning what goes on there is only now under way.

That is not to say that scientists have avoided the beach. Early in the twentieth century, people such as William Morris Davis of Harvard, who studied how sand spits grow, or Douglas Johnson of Columbia, author of one of the first texts on shore processes and shore development, began modern geology's serious study of the coast. When heavy coastal development got under way and erosion almost immediately began threatening the new buildings, these early researchers, Johnson in particular, were among those who responded to the calls for help.

Because it was the first state whose coast was substantially developed, New Jersey was the first to suffer substantial ill-effects from erosion. In 1922, the state Board of Commerce and Navigation, alarmed by the erosion threat, financed studies of where the erosion was occurring and how it might be stopped. The state turned to the federal government, the federal authorities turned to the Corps of Engineers, and the corps, in turn, looked to Johnson for advice.

He had already helped establish a Committee on Shoreline Studies at the National Research Council, and it was this committee that helped organize representatives from sixteen Atlantic and Gulf states into the American Shore and Beach Preservation Association, still an influential group. Meanwhile, the Corps of Engineers had organized its own Board on Sand Movement and Beach Erosion. Ultimately, this organization evolved into the Beach Erosion Board and then the Coastal Engineering Research Center, but when it began operating, coastal research was preliminary and imprecise. Measuring sand movement on the shoreface was the work of graduate students who carried simple surveying equipment in the surf, where they would make measurements until the water was too deep, cold, or hazardous for them to continue. The sand movement board conducted some of its research by giving students diving helmets and ordering them to sit underwater and take samples. Other researchers made measurements from ships, but ships rarely ventured close to shore. Working there was just too difficult.

This primitive research produced primitive engineering. In those days, breakwater design was simple, according to Robert Wiegel, an emeritus professor at the University of California and one of the early members of the "Berkeley Mafia" of coastal engineers. At a conference honoring him in 1993, Wiegel recalled the thinking that prevailed when he began his career in the 1930s: "You make breakwaters out of the biggest, heaviest rocks you can find and you build 'em as big as you can—that's the way to build breakwaters." Another eminent coastal scientist who got his start in that era, Willard Bascom, described his early early attempts to anchor instruments in the surfzone. "The instruments beat us back to shore," he related. "We had a lot to learn about waves in those days."

The problems of studying the geology of the surfzone are so daunting, in fact, that large-scale research did not get under way until the invasion planning of World War II forced scientists into the water. To prepare for the D-day invasion, the British collected detailed information about the geology of Normandy beaches from every possible source. Old guidebooks were ransacked for data on tides, currents, and beach topography. A radio appeal to prewar vacationers produced ten million old postcards—thirty thousand the first day. Aerial reconnaissance provided panoramic photographs. Members of the French resistance in Paris stole four volumes of geological maps from the national library in Paris, including an ancient survey by the Roman army. They smuggled them to London, where classicists translated Latin notes on coastal peat deposits.

But the invasion planners had to know more about the beaches, in particular whether heavy equipment could roll across them without bogging down. So on New Year's Eve 1943, the No. 1 Combined Operations Pilotage and Beach Reconnaissance Party, a two-man coastal research team armed with pistols, daggers, and sample tubes, set out in a midget submarine to cross the English Channel and hit the beach. In his book *D-Day* Stephen Ambrose describes what happened next: "They crawled ashore, walked inland a bit, went flat when the beam from the lighthouse swept over the beach, walked some more. They made sure to stay below the high-water mark so that their tracks would be wiped out by the tide before morning. They stuck their tubes into the sand, gathering samples and noting the location of each on underwater writing tablets they wore on their arms."[1] Burdened by their heavy samples, they headed into the surf toward their midget sub, only to be thrown back up on the beach by the breaking waves. They tried again, and the waves rebuffed them again. On their third attempt, they made it.

The teams made surveys all winter and the samples they collected showed the Normandy beaches would indeed bear the weight of invasion armor. Later, the places they surveyed were given code names: Sword, Juno, Gold, and Omaha.[2]

Meanwhile, American military planners were telling coastal scientists it was their patriotic duty to study conditions on the beach. Bascom himself got his start on wave research as part of the World War II Waves Project at Berkeley. The goal was to survey underwater topography, in order to learn what characteristics of waves and sand would cause trouble for landing craft approaching enemy-held beaches. His study area was the beaches of the Pacific Northwest. The project's director, Morrough P. O'Brien, the dean of engineering at Berkeley, an early researcher for the Board on Sand Movement and Beach Erosion, and later a member of the Beach Erosion Board, chose this stretch of coast for its roughness, Bascom recalled years later. "His theory was, 'If you can work there, you can work anywhere.' Subsequent experience certainly proved him correct."

Soon, Bascom found himself at sea in a thirty-two-foot amphibious truck, working with surveyors on shore to track changes in the bottom topography of the surfzone, as fifteen- and twenty-foot breakers routinely crashed over them. "Somehow, in innocence and ignorance, I was persuaded that fifteen-foot breakers smashing down on a thirty-two-foot tin boat were nothing to be disturbed about." he wrote later. "Perhaps when the Coast Guardsmen from Humboldt Bay lifeboat station served notice that we were working at our own risk and could not count on their help if we got into trouble, I should have been more wary. They obviously were astonished that anyone would start out into what they considered to be a raging surf for any reason short of emergency life-saving. . . . We were the first, and to this day probably the only ones foolhardy enough to take this much interest in the sand beneath the winter surf on northern Pacific beaches."[3] To make matters worse, he noted, "since the beach constantly readjusted itself, there was no end to the work."[4]

After the war ended, O'Brien, Wiegel, Bascom, and others turned to civilian pursuits, usually with an eye to engineering harbors or other structures on the rapidly developing coast. On the East Coast, a series of devastating hurricanes in the 1950s, followed by the notorious Ash Wednesday northeaster in 1962, brought pressure for more elaborate coastal engineering projects, and in 1963 the Beach Erosion Board was reorganized as a research agency, the Coastal Engineering Research Center. Its broad mandate called for research on the coast. CERC, as it was known, established its headquarters at the Waterways Experiment Station in Vicksburg, Mississippi, on the Mississippi River, in the early 1980s. In 1996 it was combined with the corps's Hydraulics Laboratory to form the Coastal and Hydraulics Laboratory.

Today, there are also major centers of coastal research at Scripps, the Woods Hole Oceanographic Institution, Berkeley, Oregon State University, the University of Florida, and Louisiana State University, to name a few. Many schools of engineering operate their own wave tanks or flumes, where machines generate scale-model waves and send them crashing on scale-model beaches, so scientists can study the results. Still, work in the surfzone is as crucial and as daunting as ever. That is why the scientists come to Duck.

When the corps was looking for a place to do coastal research, the Duck site had much to recommend it. It is on a relatively straight stretch of coast, far from inlets and their confusing currents and other one-of-a-kind features. Because the tide there rises and falls only about a foot on average, it has little influence on the wave effects the scientists want to measure. The Outer Banks are notoriously stormy, an ideal place to measure erosion events on a beach. Because the barrier island is less than a mile wide at the pier, the oceanfront site also offers easy access to Currituck Sound, for studies of sound-side erosion. Norfolk is near enough and large enough to support the researchers' transportation and equipment requirements. Best of all, the land was available. The navy had finally stopped using it as a bombing range.

The Field Research Facility (FRF) comprises two buildings with equipment rooms, labs, and offices; a tower; and the pier, twenty-five feet above mean sea level and fitted with tracks for moving heavy equipment. Birkemeier and his staff work all year collecting the data recorded by their instruments, or putting still other instruments in place for more measurements. Often, these instruments are installed in the surf on skeletal pipe frames or "sleds." These sleds are towed into surfzone by a surplus amphibious vehicle the FRF obtained from the army or by a contraption designed at the pier, the Coastal Research Amphibious Buggy, or Crab. The Crab has a small cab, barely big enough for two to stand in, mounted atop a thirty-five-foot wheeled tripod. Powered by a fifty-three-horsepower Volks-

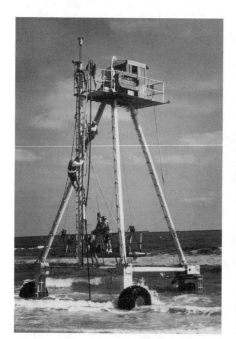

The Coastal Research Amphibious Buggy, or Crab, takes instruments into the surf for precise measurements. *(Corps of Engineers)*

wagen engine, it rolls slowly into the surf on water-filled tires, at a top speed of two miles per hour, looking like a gigantic, gangly water bird and shuddering with every wave. (Scientists in the Netherlands recently built the world's second such vehicle, and dubbed it Eurocrab in tribute.)

Strange as it looks, the Crab has been invaluable for tracking changes in the shoreface at Duck. Instruments mounted on its cab and linked by laser beam to instruments on the roof of a building on shore track the Crab's elevation so precisely they can detect as little as two centimeters of sand eroded or deposited on the bottom. Technicians in the Crab make measurements over well-established survey lines perpendicular to the shore nearly every day and especially before and after storms. When these measurements are correlated with wind, wave, and other records, the result is a precise account of how changes in the weather produce changes in the shoreface. Though accuracy within two centimeters is astoundingly precise by current standards of coastal research, Birkemeier wishes the Crab's measurements were even sharper. Two centimeters may not sound like much, he pointed out, but a two-centimeter layer of sand over miles of beach is a lot of sand.

From the scientists' perspective, one of the best things about the Duck pier is that the staff takes measurements often and over the long term, so scientists can tease out the trends that can be swamped by the anomalous effects of a single stormy winter or even a single storm. By now, the pier has years of data on weather, wave direction, width of the surfzone, water density, bottom configurations, and other factors.

"It's tough to go to any beach, stay for a short time, and make profound statements," Birkemeier said. "The beach does tend to erode in storms and rebuild in nonstorms, but each occurrence is unique because you have different starting points." Scientists now know that the biggest changes to the Duck beach occur in September and October, after months of calm summer swells have carried sand onshore and the beach is at its widest—and most unstable. When the storm season begins then, the beach is poised to erode. By February, the beach is at its narrowest, having responded to months of heavy weather by shifting its sand into bars offshore. "In February, a storm has to be something really big to really rearrange things," Birkemeier said.

Scientists rely on the pier's baseline data when they descend on Duck in periodic bursts of experimentation. In turn, they add their data to the mix. Someone such as Guza, who might be studying the effects of bubbles generated by waves, can correlate his data with that of other researchers studying currents or the movement of sediment in the water. The group

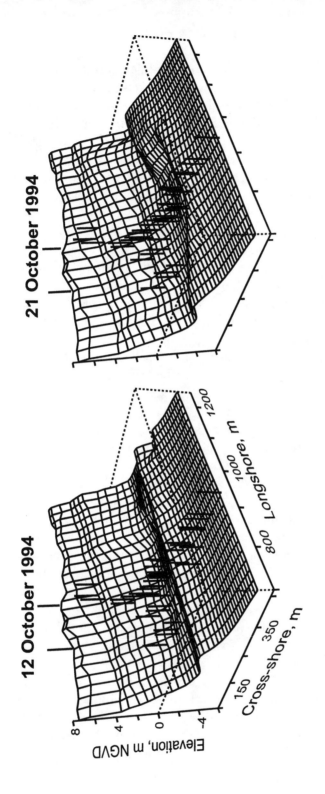

During SandyDuck, a research effort organized by the Corps of Engineers, scientists working at Duck, N.C., used specially made instruments to map changes in the shape of the beach, both above and below the water line. These graphs, made nine days apart, show how the sandbar changed in a storm. The small lines sticking up from the grid represent other instruments arrayed in the surf and on the beach to measure weather conditions, currents, wave height, and the like. *(U.S. Army Corps of Engineers Waterways Experiment Station Field Research Facility, Duck, N.C.)*

efforts are also desirable because of the extreme difficulty of maintaining instruments in the water. The worse the weather is, the more important it is to keep the instruments running, something that the scientists can only achieve by helping each other.

Their experiments generate so much data—a single instrument operating continuously may generate millions of data points—that it takes the scientists years to analyze, review, and publish them in research journals. "We spend two months on the beach and the next four years in front of computers figuring out what it was we saw," Peter Howd, a coastal geologist then at Duke University, grumbled during a 1994 experiment. In turn, these results generate new questions, calling for still more research. So even as one experiment is under way, Birkemeier and the scientists are thinking about the next one.

The first experiment at Duck took place in 1982, and it provided the first-ever measurements of how the bottom changes in the surfzone during storms. But the surveys were cumbersome and the results lacked detail. A similar, larger, experiment in 1985 provided more detail and tracked, for the first time, how sandbars form, migrate, and change shape during and after storms. The pier staff took the Crab into the surf along nineteen survey lines, covering almost five hundred yards of of beach, up to six times a day for seventeen days, before, during, and after a northeaster. (In their paper reporting the work, Birkemeier and Howd noted that it would have been impossible without the fortitude of the men who took the Crab into the surf at "all hours of the day in adverse weather conditions.")[5]

Unfortunately, the new information only pointed to how much coastal geologists did not know about currents in the surfzone, an impression reinforced in the next experiment, "Superduck," in 1986. Superduck involved thirty research projects, concentrating on longshore currents in the surfzone. In studying certain of these currents, called edge waves, researchers stumbled on something they had never seen before, a kind of instability or oscillation in the longshore current. They called these oscillations "shear waves." Shear waves had never been observed or even theorized about before, but coastal geologists now believe they are "an important missing link to our basic understanding of nearshore dynamics," as one of their discoverers, Joan Oltman-Shay, a senior scientist at the consulting firm Quest Integrated, put it. Though scientists still do not know exactly why they occur or how they affect the beach, their discovery showed that longshore currents are more complex than they had thought.

Edge waves themselves are a forceful shaper of the beach. Unlike the everyday waves that come crashing in on the shore, edge waves travel parallel to the beach. And unlike ordinary waves, which are generated by wind blowing across the surface of the ocean, edge waves are generated in the complex sloshing that goes on amid the deceptively simple breaking, uprush, and backwash of the waves on the beach.

The distance from crest to crest of the edge waves is long, so their frequencies are much lower than ordinary wave frequencies, a matter of minutes, not seconds. Their height—the distance from the crest to the trough of each wave—is a few centimeters at the most, so small that the waves are invisible to the eye. People on the beach experience an edge wave, though, every time they get their feet wet when a wave comes unusually far ashore. Surfers, who recognized this phenomenon as early as anyone, call it "surf beat."

Mathematicians first began theorizing about edge waves in the midnineteenth century. One hundred years later, in 1949, Walter Munk of Scripps first measured them on California beaches. Later, he and others tried to generate them by sending ships up and down the coast at high speed. But scientists are only now beginning to discover how important edge waves are. Recent research suggests they have a big influence on the amount of sand carried along the shore and in the formation of crescent-shaped sandbars, rip currents, and beach cusps, the regular undulations that form on most beaches as if by magic—and in ways still unknown.

Edge waves also help explain a coastal paradox: storm waves have the most energy where they break and the least where they run up on the shore, yet they produce tremendous erosion on shore. Edge waves, which appear to increase in amplitude the closer they get to shore, may be part of the answer. The details of edge waves have not even been identified, let alone analyzed, but they are attractive to coastal geologists because they explain many nearshore features simply and quantitatively "without having to invoke a liberal degree of arm waving," says Rob Holman, a coastal scientist at Oregon State University who has led several research projects at Duck.

From a policy standpoint, the crucial thing about edge waves is this: fifty years ago they were practically ignored, and even today scientists are only beginning to measure their components and actions. More than anything, these waves prove how much about the coast remains unknown.

The work at Superduck was rewarding but exhausting. "We came out of that with every piece of equipment broken or worn out or destroyed and every member of the staff broken, worn out, or destroyed," recalled Carl Miller, an oceanographer at the pier.

The 1990 experiment, dubbed "Samson and Delilah" (Samson for "Sources of Ambient Micro-Seismic Ocean Noise," a project Guza was running for the navy, and Delilah for Samson), concentrated again on waves and currents. In 1994, the emphasis was again on water motion but also on sediment and the effects sandbars and other beach "structures" have on waves and currents. More than one hundred scientists—from universities in the United States, Britain, and Canada; consulting firms; the navy's research laboratory and postgraduate school; Woods Hole and Scripps—descended on the pier to test new instruments they had devised to measure the movement of sand under the surf. They installed arrays of equipment in the surf in summer, when the weather was relatively calm, and returned in September, when it began to get stormy, to measure currents and sediment. The experiment was long and complicated, but it was only a run-through—a shakedown of instruments that were used in the next big experiment, "SandyDuck," in 1997.

Fortunately for the perennially cash-poor researchers, the military seems as interested in the beach today as it was in World War II. The research, while answering the scientists' questions about erosion, will also give the navy useful information for planning amphibious operations. But Birkemeier is worried that federal budget cuts, particularly efforts to get the Corps of Engineers out of the beach nourishment business, could rob his research station of funds, especially if more coastal scientists start to rely on numerical models rather than real-life observation of the coast.

In part, the scientists who do research at Duck are working to provide the data needed to make accurate mathematical models of the coast. But they know perhaps better than anyone else how difficult that will be, and they are the first to admit that it is impossible to apply, en masse, findings made at one beach to another. Still, given the rigors of research in the field, accurate models would be a tremendous boon to scientists and policy makers. In the past, researchers did not have the computer power necessary to analyze the kind of data generated by an experiment at Duck, where a single researcher—Guza, for example—might have sixty sensors in the water, each generating a data point four times a second, day and night for weeks. But now they do, and they are adapting the new computer power and the setup at Duck to the standard processes of scientific research: observation, hypothesis, experiment, and evaluation.

   Much of their research lies in three areas: where, how, and when sand moves; how can scientists use existing data to predict shoreline change, particularly erosion rates; and, what are the effects of shoreline structures, especially the continuing parade of systems designed to prevent erosion. To solve these problems, scientists pose many questions. To name only a few:

- Does the beach have an "equilibrium profile," a slope it strives to maintain by shifting its sand above and below the tide line? For years, this equilibrium was axiomatic among coastal researchers, who declared it could be determined by analyzing the grain size of sand on the beach and the prevailing wave patterns. Research at Duck and elsewhere has called it into question.
- Is there a "closure" point on the offshore beach floor, beyond which sand does not move? Also axiomatic, it too has come under challenge recently. Advocates of beach nourishment have argued for years that sand that moves off the beach is not lost from the littoral system because it is trapped within the closure depth. Today, as Howd puts it, the idea of closure depth may not be dead, "but neither is it alive." If the closure depth theory is discredited, repercussions for beach nourishment could be severe.
- In addition to moving along the coast suspended in the littoral drift, how does sand move along the bottom in ripples or other "bedload formations" that creep along the coast? Which transport mechanism is more important? Under what circumstances?
- What about other kinds of low-frequency edge waves or infragravity waves? Guza, who thought he had answered the beach cusp question years ago with an explanation drawn from work on edge waves, says work drawing on theories of chaos and complexity theory has reopened the issue. When other researchers studied the phenomenon at the Canaveral National Seashore in Florida, where cusps were spaced about twenty-eight yards apart, the data they gathered supported either approach.[6] "Overcoming this will be difficult," they said.
- Did barrier islands form from erosion of coastal terraces in an earlier geological era or, as some scientists now maintain, are they formed of sand carried in by storms?
- How, when, and why do sandbars form and migrate?

The year-round work at Duck has provided startling new information about the behavior of sandbars. Until fifteen or twenty years ago, researchers thought most beaches were relatively flat in summer, and changed in winter to a steeper slope with one or more offshore bars. After expanding the research he began at Duck, Holman discovered that this is not necessarily the case. "We found out the whole system is a much more dynamic system than we thought," Holman said. "And the reason we didn't know before was we didn't sample more than twice a year."

He often cites one of his graduate students who studied sandbars at Duck, where in winter the beach normally has inner and outer bars. For his master's thesis, this student created a mathematical model to explain the sand movement. When he returned for work on his doctorate, however, the outer bar was gone. "The whole situation had changed," Holman said. "I have no idea why. Not a clue. Plenty of stuff about bar formation and behavior is well beyond our ability to explain. But if we didn't have long-term observation, we wouldn't know it had happened. We would be modeling the wrong thing."

Holman is engaged in a long-term effort to find out just how complex sandbar behavior is and to devise an easy way of keeping track of it. His tools are video cameras mounted in high places overlooking the beach, like the metal tower at Duck or a bluff at Yaquina Head, Oregon, and computers that analyze the photos they take every ten seconds, as long as there is daylight. The cameras record the waves breaking on the sandbar and then on the beach. By comparing the tapes made at Duck with the meticulous data from the Crab, Holman was able to devise computer software to digitize and combine the real-time images into ten-second averages in which the apparently roiling surf resolves itself into clear patterns of waves breaking over sandbars. Over time, these patterns reveal the appearance, migration, and disappearance of sandbars. As the computer looks beneath the cacophony of the breaking waves, it can see outer bar formations, roving troughs of deep water, and other features invisible to the eye under the mish-mash of tumbling breakers. Among other things, the work suggests that second, outer bars form far more often than scientists had believed, and that their influence may be important.

The computer/camera system, which Holman dubbed Argus, has now also been installed at beaches in Hawaii, Australia, and the Netherlands, and the pictures are posted on the Internet, almost as soon as they are produced.

Today, scientists seeking to predict shoreline erosion can consult surprisingly a large amount of historical data in the form of maps, charts, and aer-

A camera mounted on a metal tower at the corps research station at Duck photographs the beach every ten minutes. The images, combined by a computer, sort out the confusing jumble of breakers into a coherent picture of sandbars as they form and move. *(Rob Holman)*

ial photographs. Now they are developing new techniques to make this data usable. It is not an easy process.

Accurate surveys were done on the East Coast starting in the mid to late 1800s, but nowadays scientists rely more on aerial photos, which first appeared in large numbers seventy years ago. New computer techniques allow researchers to correct for distortions from camera angles and digitize the results for easy comparison. But there are still plenty of problems. For example, it is impossible to tell from an aerial photograph where the mean high water line is on a given beach. The photograph will show only the high water line at the time the picture was made—or the previous high tide. Photos of low-lying sandy beaches are especially prone to error, because differences in water level are exaggerated there, even if the beach itself has not changed much. Also, seasonal changes in the beach, which may be large, can mask smaller, incremental changes going on over a period of years.

Theoretically, the longer the time period covered by the data, the more reliable the conclusions drawn from that data will be. But even a long string of data can create erroneous impressions. For example, a coastline that oscillates back and forth, as many do, may appear in the short run to be accreting, only to start eroding again. Without long-term data, it may be impossible to know whether observed changes at a beach are part of a long-term trend or merely short-term phenomena. Unfortunately, though, coastal policy is often based on small, possibly unreliable sets of data. Everyone agrees policy makers need a steady stream of reliable information, but few of them get it. That is because this kind of data requires monitoring, and monitoring is the unattractive stepchild of scientific research.

It is not glamorous to wade into rough surf and freezing water and repeat the same measurements week after week, month after month, year after year. And it's not cheap, either. It may take years to tease meaningful trend lines or other findings from the mountain of data, too long a time frame for most coastal projects. And once a project is built, few government agencies want to spend money simply watching it.

"Real scientists don't do applied wave research because they would rather sit in their labs," said Gary Griggs, the seawall expert at UC Santa Cruz. "It's the number crunchers versus the watchers," he said, "and the watchers are out of style."

But until much more watching is done, all models will be incomplete, imprecise, or downright wrong. Some coastal geologists believe beach conditions are so site-specific and depend on so many variables that change from place to place, that modeling will never offer useful infor-

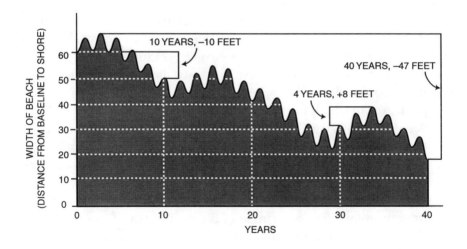

Beaches gain and lose sand constantly, from storm to storm, season to season, year to year, making it difficult to estimate what underlying trends are. Over the long term, the beach shown here seems to be losing about 1.2 feet per year. Observed for one 10-year period, though, it seems to be eroding at a somewhat slower pace. Measured at another, 4-year period, it seems to be gaining sand, growing by 2 feet a year. *(Jen Christiansen)*

mation. On top of that, modelers are only now getting around to working storms into their equations, and modeling storms is as difficult as modeling beaches.

Even some engineers are worried about the proliferation of computerized models, given the lack of data about the processes that take place on the beach. As Birkemeier put it, "It doesn't matter how fast the computers are if the physics are wrong." He said, "We would like to do realistic models of beach erosion. We would like to take a fine-grain beach and move every grain of sand around for twenty years. Obviously we can't do that." Though he believes that modelers will eventually succeed, he added, "It's probably easier to go to the moon than to understand the physics of the nearshore."

It will take monitoring to answer one of the most vexing and persistent questions in coastal engineering: do seawalls exacerbate erosion? Research at Duck cannot answer this question, because the site was chosen specifically because there was no coastal armor nearby. The answer can come only from extensive, widespread monitoring, something the corps has only recently started to finance.

Unlike the so-called placement loss that occurs when a wall impounds sand behind it, or the "passive erosion" that occurs when encroaching seas encounter an immovable wall, "active erosion" occurs, in theory, because the wall interferes with normal wave processes such that waves scour the bottom in front of the wall.

The degree to which this active erosion occurs is the subject of lively debate. Some coastal geologists fear it is a big problem wherever there is a seawall. Others are not so sure. One of the unconvinced is Griggs. Griggs became interested in the question in the winter of 1982–3, when he got a frantic call from the owners of an apartment house on Del Monte Beach, in Monterey Bay, California. It was an El Niño winter, a season of unusually fierce storms, and the owners feared erosion would send their complex crashing into the sea. They were seeking state permission to protect the property with a rock revetment and they needed a scientist to back their assertion that it would not harm the beach.

Griggs and his colleagues had done some work on the performance of coastal structures such as seawalls, revetments, and the like, but that work had dealt strictly with whether this armor protects buildings. In Del Monte Beach he confronted a different question: Does it damage the beach?

"I started looking into information," he related. "There was almost nothing I could find. There seemed to be this kind of intuitive sense that if you build a wall you reflect wave energy and therefore you scour the beach. I found out it was really hard to say. So I was kind of at a loss — they should do this, they shouldn't."

In the event, he concluded that because the proposed structure was a revetment of loose rock rather than a solid wall it would have relatively little impact. (The revetment was built, but later the state ordered it removed — it has been built on a public beach.) But the experience had started him down a new research path, an effort that acquired new urgency as more and more Californians, frightened by the storms of the 1982–3 winter, sought permission to build seawalls. Like other geologists, he had more or less presumed that building a seawall would accelerate beach loss. "If you put a vertical, impermeable concrete wall out on a beach, it stands to reason it will reflect energy more than a beach, that it would have some obvious effects," he said. This impression is supported by experiments in wave tanks in which scientists build model beaches, create waves, and watch what happens. If model walls are added to the wave tank beaches, scour patterns quickly form in front of them, as waves bouncing off the walls carry sediment away.

Unfortunately, though, whether the questions involve seawalls or sandbars, it is difficult to answer them in a wave tank, where it is impossible to re-create the conditions of a natural beach. While natural waves vary, experiments in a wave tank usually involve waves of a fixed height and frequency — what scientists call a "uniform wave regime." Unlike the real beach, the model beach has sand of uniform grain size — and since sand grains would be too large and heavy for a scaled-down model beach, tank operators must make do with alternatives, such as ground coal. Plus, almost any natural beach is far more irregular than any beach built in a wave tank. The tank models cannot account for longshore currents, inlets and their effects, changes in water table, and so on.

"You can't make eight-foot waves on a beach that's six hundred feet long in a wave tank," Griggs said. "You really have to measure what's happening out there."

So, in research financed in part by the Corps of Engineers, he chose several beaches in Monterey Bay, some with walls, some without, where he would make careful measurements over a period of years and record what happened. One of his sites was a seawall in Aptos, a town clinging to the bluffs on the coast south of Santa Cruz. The seawall there was a con-

crete structure about ten feet high, running along the base of a bluff tow-
ering eighty feet or more above it. The wall is designed, as most are, with
its concave surface facing the waves, to absorb their force. Rock rip-rap
lines its base. Modern, low-slung houses, in the million-dollar range, sit
cheek-by-jowl behind the wall, which is hung with signs warning people
to keep off. It's clear that this stretch of coast is retreating; there are fresh
slides at the top in areas with no seawalls, and people have hung netting
or black plastic sheets at the top to prevent slumping.

Griggs and his graduate students started measuring the beach in the fall
of 1986, at first every two weeks, then monthly. They found the normal
winter narrowing began earlier in front of the wall, possibly because the
sand was wetter there, making it easier to carry away in suspension. Still,
to his surprise, Griggs discovered that walled and unwalled beaches ended
up the same width each winter and widened again to the same width the
following summer. As Griggs put it, the berm appeared to have "no memory of what happened the previous winter.

"I frankly expected to find enhanced erosion," he said. "What happened was different from what I expected."

Griggs's preliminary findings caused something of a stir when they were
published in 1994. Coastal geologists opposed to hard stabilization worried that his site-specific findings would be used as ammunition in battles
to overturn restrictions on seawalls elsewhere, even though Griggs was
careful to note that his findings applied only "in this particular study, this
particular site, this particular time period." He was among those who said
the project would not be conclusive until he and his colleagues observed
the effects on the beach of a severe winter season. They got their wish, so
to speak, in early 1995, when the study area was struck with the worst series of winter storms since the infamous winter of 1983.

Though the area in front of the wall appeared to erode more deeply
than control beaches nearby, and though there was slightly enhanced erosion at the ends of the wall, these differences did not last long after the
storms had passed. "The similar responses of the control and seawall
beaches to the storm waves of 1995 were consistent with our long-term observations," Griggs and his colleagues wrote. The beach in front of the
wall "quickly lost the imprint of accelerated scour and a general alongshore homogeneity began evolving within months of the 1995 storms."
They concluded, "there is no evidence of impaired recovery."[7]

Initially at least, these findings seemed to confirm a theory a number
of coastal scientists have voiced but have been unable so far to prove:

that waves *do* scour beaches in front of seawalls more than they do un-armored beaches, but that the scoured areas are quickly repaired, as long as an adequate amount of sand is moving alongshore in the litto-ral drift.

Inman, for one, voiced this theory at the San Francisco conference hon-oring Wiegel. "The maximum scour might occur at the height of the storm, but it might be filled in in one tidal cycle and we never capture it," he speculated. "If it's that short-lived, then it has no long-term impact." If Inman is correct, though, it may mean that walls like the one Griggs stud-ies have minimal effects on the beaches they front only if nearby stretches of beach continue to contribute sediment to the littoral drift. Then the question would be: Are the walled beaches trapping sand that would other-wise have moved downstream?

The Monterey Bay research was one of two projects sponsored by the corps to monitor sea walls. The other, led by Professor David R. Basco of Old Dominion University in Virginia, was at Sandbridge, Virginia, the tiny community where residents built their own seawalls out of sheet pil-ing and other materials. For several years, Basco and his graduate students, assisted by professional engineers, studied the beach and underwater topography to a depth of twenty-five to thirty feet. They found that the loss of dry beach was statistically the same for walled and unwalled portions of the beach. But they found that the winter narrowing was exaggerated slightly at the walled portions, although they recovered almost as fully as the unwalled sections each summer.

The group concluded that the problem at Sandbridge was one of topography: steep slope, increased wave size, erosion. The walls them-selves did not hasten the process. But they saw several problems that did not appear at Monterey Bay. There was significant scour at the ends of walls. They had to measure beach width from the low water, not from the high water line, as is usually done, because there was no beach at high tide.

Here, as at Griggs's site in California, the situation might be worse but for the sand moving in the littoral drift, which might be taking the place of sand lost in front of the wall. And at Sandbridge, scientists estimate, the drift carries a lot of sand—perhaps three hundred thousand cubic yards a year.

Few engineering disciplines are as complicated as coastal engineering. As Wiegel says, "Coastal engineering is one of the most difficult branches of civil engineering, as the environmental loading on structures by waves, currents, and winds is always dynamic, as is the geomorphology of the beaches and nearshore regions. Furthermore, there are substantial scale effects so that full-scale field experiments are needed."[8] Like other coastal scientists and engineers, Wiegel wishes local, state, and federal officials knew more about just how complicated coastal engineering really is. Increasing their understanding of what is known and what is unknown, what is possible and what is impossible, is "urgent," he says.

But, as more than one coastal scientist has acknowledged with regret, public officials in coastal areas often have too much of their political capital tied up in coastal development. They don't want to know about long-term erosion rates. They don't want to hear that their newly rebuilt beach will be gone in a year or two. They are, as Leatherman puts it, "a constituency of ignorance."

When the National Academy convened the meeting of coast experts at Woods Hole, it was clear that science was far ahead of policy at the coast. But science takes a long view; the political horizon is close-in. "If the people can find a way to fix things for thirty years or even for the next fifteen years, they'll find a way to do so," said Don Bryant, who dealt with the problem when he was mayor of Nags Head, North Carolina, in the late 1980s. "People don't take the geologist's view."

Parkinson, the Floridian who serves on the board of an inlet district, agrees. "People are not interested in studies," he said. "They want sand on the beaches."

Coastal engineering is a young field. Crucial discoveries—like barrier island migration and the existence of shear waves—are still being made. "It's trial and error right now," Parkinson said. "People can't believe that scientists and engineers are debating each other, and experts are there and they don't agree. There's no black and white. We don't really know. We need twenty years and $20 billion and then we'll have your answers."

But evidence is growing that simply throwing money and coastal works onto the beach can actually do more harm than good. The Corps of Engineers is acknowledging that, in a way, as it revises its *Shore Protection Manual*. The corps's first guide to coastal works, *Technical Report 4, Shore Planning and Design*, appeared in 1951. It was expanded to three times its size in 1973, when it was issued as the *Shore Protection Manual*. The two-volume work was written for corps officials and district offices, but today it is in use at universities and engineering firms around the world. It is widely regarded as the most authoritative coastal engineering guide in existence.

The manual was updated in 1977 and again in 1983, but it still relied on the technology of the early 1970s—before the availability of computer systems, numerical simulations, and new information on waves and sediment transport.

Now the corps is replacing the manual with a coastal construction manual officials hope will offer expanded information and new emphasis on the environmental impact of coastal works. It includes a section on coastal geology. It will also include material on what the engineers call "coastal engineering for environmental enhancement" and on how to mitigate the evil effects of earlier coastal engineering. For the first time, it will be available on CD-ROM, so it can be updated regularly.

It's 1994, on another chill October morning. Once again the scientists have gathered in the dunes at Duck. Once again, they meet in a makeshift conference room to report on the successes and problems of the day before. As in 1990, the weather is cold, gray, and stormy. This time, a northeaster is bearing down on the Outer Banks. Heavy waves are breaking on the beach, and Holman's video system has picked up some puzzling changes in the sandbar offshore. "We don't know what they are," he said. In a concession to the heavy weather, Guza has donned sweatpants and a flannel shirt, but still his feet are bare.

Overall, the scientists report, the flow of data is good, and instruments seem to be working in concert. There are only two major problems, and there is nothing anyone can do about them. One researcher is having trouble with a large school of fish that has taken to hanging around his instrument array, interfering with some of the measurements. Another researcher, who is using radar to scan the ocean surface to measure surface currents, had set up one of his antennas on a vacant lot near the pier, "and yesterday we got a call that they are coming to level the lot because it is time to build!"

In only a few years the pressures of development have brought many complications to the work at Duck. This year, there were loud local protests when Birkemeier restricted beach access at the pier to accommodate the researchers. After a complaint from a congressional office, he opened a pathway near the dune line for strollers. "We are concerned about our ability to do science," he said. "And it implies it will be difficult to do similar science elsewhere."

Meanwhile, though, the number of scientists who want to work at Duck is growing fast. There are many more people on hand for this project than for any of the earlier ones. "I don't know how many, we're trying to keep track of them," Birkemeier says. "There are 110 people on the list, but we don't know at any given time who the hell's here. There are about thirty experiments, approximately forty principal investigators. There are graduate students. There are electronic guys. Scripps came with a welder. They all travel with an entourage of expertise. We really work to make sure they leave happy."

He said Canadian researchers arrived with $711,000 worth of instruments and three aluminum frames to hold them in the surf. "Waves jacked two of the frames out. We repaired them. We didn't want to see those long faces." The Canadians set up shop in yet another trailer nestled in the dunes, flying the Maple Leaf flag. ("Every time it gets cold, we are blamed," one of them complained.) Nearby was Guza's trailer—flying the Jolly Roger. It had taken two weeks of work the previous July to install Guza's instruments, but the work was worth it. "When the beach changes, he knows," Birkemeier said.

At midday, many of the researchers stop work and gather on the windy beach to watch the day's main event: an attempt to tow another sledfull of instruments in the surfzone. Only the Crab could do this job, and today even the Crab is pushing the envelope. It is designed to withstand six-foot waves, and they are at least that big. The Crab shudders violently with each of them. To make matters worse, the presence of the sled means it will not be possible to winch the Crab back on shore if something goes wrong. But everything goes as planned, and at the next morning in the trailer researchers report instruments on the sled are churning out data.

"It was blowing thirty-five this morning, and it's all of that now," Birkemeier says contentedly, looking out his office window later at the waves crashing on the pier. "This is just the kind of weather we need."

Three years later, this run-through pays off when the full-scale experiment, SandyDuck, begins. Problems with instruments have been ironed out or designed away. Computer algorithms for acquiring and storing data have been made more efficient. Gaps in the research agenda have been filled in. For one thing, the scientists realized that they would have excellent records of sand deposition in storms, but much less detailed data of what happened to that sand after the storm ended. So they made provisions for taking core samples immediately after weather "events" to track the sand more closely. An elaborate schedule for installing, maintaining, and removing instruments has been drawn up, almost down to the hour.

After Duck94, the researchers also resolved to pay more attention to the fields of ripples that form on the bottom. These ripple fields are emerging as a major factor in the movement of sand along the coast, but much about them remains a mystery. "We did not appreciate how complex ripple fields could be," Guza said one morning as he walked the beach and looked out at the instrument arrays sunk in the surf. "Ripple crests merge, they twist, they rotate, they come, they go." He paused, as if pondering the way new information has once again complicated the life of the scientist. "In a way, ignorance was bliss."

A few yards down the beach, a few men were hunkered down around a metal-pipe frame about seven feet square, fiddling with the dozens of sensors, meters, and other devices mounted on it. Just as they finish their fine-tuning the Crab slowly rolls into position. They chain the metal sled to the Crab and watch the Crab slowly pull it into the surf until most of the framework has vanished under water. Another day of research has begun.

Planning for this experiment began in 1991, and it will probably be 2000, at least, before the data collected have been analyzed and reported in scientific journals. By then, with luck, the scientists of SandyDuck, and other researchers, will have harvested still more data.

By now, Birkemeier has seen people who first came to the pier as graduate students mature into major figures in coastal research. "We have seen all these kids grow up," he said. More significant, "In the past few years we have seen a coming together of the process folks with the geological folks. Generally these groups are on different time scales. But there are some interactions going on here that are not inconsequential."

Among other things, the scientists are discovering that the quality of the material that underlies the beach—river mud, bedrock, or whatever it may be—seems to affect the ways the beach responds to changing conditions. "Sand over fluvial mud moves differently than sand over more solid material," Birkemeier said. "Sediments that have a lot of biological material in them erode differently than sand that doesn't." Also, coastal geologists increasingly realize that to answer questions about the ancient beaches preserved in earth's rock record it is necessary to study what happens at a place like Duck today.

These insights add still more layers of complication to the study of the shoreline. "The more details you look at, the more you have to question what your basic assumptions are," Birkemeier said. "We never had the wherewithal to even ask these questions before."

CHAPTER NINE

Constituency of Ignorance

> I cannot direct the wind, but I can adjust my sails.
> —Message seen on a church signboard in Harlingen,
>   Texas, by Bob Sheets, then head of
>   the National Hurricane Center

The Cape Hatteras Lighthouse, a slim brick tower painted with broad diagonal stripes of white and black, rises 208 feet above the dunes on the outermost stretch of the Outer Banks of North Carolina. It was built in 1870 to warn ships away from the Cape's ever-shifting shoals, "the graveyard of the Atlantic." But it was more than a beacon for mariners. People still live on the banks today who can recall the time when travel was not by road, but by horse cart on the beach, and the foggy nights when they urged their horses forward, despairing of home, until the reassuring lighthouse beam reached them through the mist.

For the rest of the state, the light is a symbol of a more romantic past, a link to a maritime tradition threatened by modern ways. And though there are other, smaller lighthouses on the Banks, none has as firm a grip on the hearts of North Carolinians as the stately tower that stood until recently at the very edge of the sea.

When it was built, the lighthouse was more than sixteen hundred feet inland, safe from the ocean's grasp. But of all unstable coastal landforms, capes are among the most unstable—and Cape Hatteras is more unstable than most. Over the years it has changed shape and position dramatically, but always retreating, retreating landward. By the time of World War I, not even fifty years after it was built, the lighthouse had lost eleven hundred feet of protective beach, and by the end of World War II, with erosion continuing, the federal government had removed it from active service, replacing it with a metal tower further inland. Soon after, the old brick tower was deeded over to the National Park Service,[1] under whose operation it immediately became one of the state's major tourist attractions.

But its position was ever more precarious. During the war, the navy built a small base north of the light and installed three groins to trap sand in front of it. Erosion worsened downdrift at the lighthouse. The downdrift groin was strengthened, which helped somewhat, but waves soon began cutting around its southern flank, threatening the lighthouse again.

North Carolinians started worrying, and their fears intensified in 1982, when a fierce winter storm brought waves to the lighthouse base. By then, the park service had officially adopted a policy against armoring the coast, but when the storm struck, workers rallied to save the light, exploiting a parking lot project underway there. "Park workers tore up the parking lot while the storm was raging, which was quick thinking, and threw pieces into the sea" around its base, said Pilkey. (The slabs of asphalt remain at the base of the lighthouse.) The beach recovered a bit after that storm, but it was clear to the park service and virtually everyone else that the lighthouse was doomed unless something major was done to save it. But what?

A businessman from the western part of the state established a fund to purchase what was then the latest thing in shoreline protection technology: artificial seaweed. The idea was to re-create the natural kelp beds that seem to protect some California beaches from erosion. The artificial fronds of seaweed, weighted with sandbags to keep one end at the bottom, would slow the movement of water, causing sand to drop out and settle in the nearshore in front of the lighthouse. Schoolchildren around the state contributed their nickels and dimes to the project and the seaweed was installed, but the project quickly failed. Fronds broke loose and caught in the propellers of passing boats. Others ended up in tangles on the beach. It was a mess.

The beach in front of the lighthouse had already been renourished several times in the 1960s and 1970s, but erosion more than kept pace with the engineers. "One of the smaller projects is famous in the annals of beach replenishment history," Pilkey related. "The pipe was still there, they were just finishing the pumping project, and a storm came in and took away the sand and all the pipe."

Desperate, the park service approached the Corps of Engineers about protecting the lighthouse with a seawall, so that it would survive even if the land around it eroded away. In effect, the lighthouse would go to sea, like the Morris Island Lighthouse, two hundred miles to the south in

South Carolina. But the Morris Island Lighthouse has a heavy foundation, on supports driven deep into the earth. By contrast, the Cape Hatteras Lighthouse, a twenty-eight-hundred-ton structure, sat on a foundation less than seven feet deep. Its base rested, in turn, on giant yellow pine beams whose strength is legendary—as long as they remain submerged in fresh water. Fresh water is rarely far below the surface on barrier islands, but if the island eroded away from the light, the fresh water would go with it. In short, if the Cape Hatteras light went to sea the way the Morris Island beacon did, its foundations would fail and it would collapse.

"It will fall," Pilkey predicted. "It will simply topple over."

Nevertheless, the corps pressed ahead with seawall plans even though these called for a wall so high it would hide much of the lighthouse, especially its distinctive base. Some engineers also speculated that pumping enough water to mix the necessary concrete would lower the nearby water table enough to threaten the yellow pine logs. And even at that, the lighthouse probably would not survive.

So Pilkey and several others interested in the fate of the coast took another approach. If the lighthouse cannot survive where it is, they reasoned, it must be moved. In the process, it would become a powerful symbol for what Pilkey called the only sensible approach to accelerating erosion: retreat from the beach. He and a group of like-minded people formed the Move the Lighthouse Committee, which "made quite a bit of noise." As a consequence, the park service asked the National Academy of Engineering, a congressionally chartered organization of eminent engineers, to look into moving the lighthouse. The academy concluded in 1988 that moving was not only feasible but also the most reliable method of saving the lighthouse. In its report, the academy said the lighthouse could be moved on rails laid on a bed of concrete designed to deteriorate once the job was done. The park service endorsed the idea a year later.

Though many people feared the lighthouse would never survive a move, in fact it was well suited to moving. Its center of gravity is low in its red brick and gray stone base, and if the tower itself were adequately braced from within, there would be little danger of collapse. "Off-the-shelf technology," Pilkey called it. Plus, there was plenty of room to move it. The report designated a spot about a quarter of a mile inland, where it said the lighthouse should be safe for generations.

But opposition rose immediately. Some who loved the lighthouse distrusted the engineers' assurances that it could withstand the rigors of a move. Others objected that their beloved lighthouse would look ridiculous jacked up on blocks like a derelict car. They said they would rather see it fall to the ocean, if that was its fate, than have people mess with it. This, too, was fine with Pilkey, because it fit right in with his philosophy of letting nature take its course on the coast. "If you can let that lighthouse fall, you can certainly let a condominium fall," he declared. "It would set a beautiful example of coastal management for the rest of the country."

That was exactly what bothered the people who had homes and businesses near the lighthouse and elsewhere on the coast. Like Pilkey, they recognized the lighthouse's power as a symbol. If the government decid-

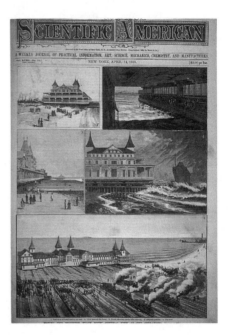

Left: When erosion began threatening the Cape Hatteras Lighthouse, the National Research Council recommended moving it out of harm's way. A citizen's group, the Move the Lighthouse Committee, promoted the idea with posters like this one, including a drawing from the *Washington Post*. *(Orrin H. Pilkey, Jr.)*

Right: Moving buildings is not new. Nor is it necessarily perilous. In its issue of April 14, 1888, *Scientific American* reported the moving of the Brighton Beach Hotel in Brooklyn. *(Scientific American)*

ed *it* should not be maintained in place, surely they could not expect much help if their shops and motels and condos were threatened. Opposition among commercial interests along the coast began to grow.

By the early 1990s, the park service had promised to begin the elaborate preparations for moving the lighthouse as soon as it was in imminent danger, but meanwhile it installed sandbags over the remains of the parking lot asphalt. Advocates of the move smelled a rat. The lighthouse was already in imminent danger, they said, and the next big storm could take it out overnight. Once it was about to topple, it would be far to late to clear the new site, lay tracks, and make other necessary preparations. "The right storm from the right direction will tomorrow take this lighthouse," Pilkey told a group of students on a field trip he led to the Banks in 1991. He and the rest of the committee fear advocates of armoring the coast are urging delay in hopes that when the lighthouse starts to fall, the government will rush in with emergency armor to protect it. Then, they will use this armor as a precedent for other projects to protect their property and, inch by inch, some of the nation's most spectacular beaches will be lost.

In 1996, the North Carolina Division of Coastal Management told the park service it could protect the lighthouse base with three hundred new sandbags and it could hold in place for five years. Then, presumably, it would move the lighthouse. Many who love the lighthouse worried in the late summer of 1998, when Hurricane Bonnie took aim at the North Carolina coast. Ironically, though, it seems the storm's swells added rather than removed sand from the lighthouse base. Meanwhile, plans for another groin are pending, but work has begun on the move.

There are three approaches to coastal erosion: armor, beach nourishment, and retreat. Until the twentieth century, retreat was the favored option. A growing number of people say it should be our policy again.

Many spots on the coast, particularly on the East Coast, offer silent testimony to the people who settled close to the beach, only to abandon their homes or farms once they realized how unfit coastal land was for settlement. Hog Island, Virginia, for example, was settled in the eighteenth century. But sand comes and goes there in dramatic shifts researchers are only now beginning to track. The last residents, worn out by their struggle with the landscape, left Hog Island early in the twentieth century. Today, the island is part of the Nature Conservancy's Virginia Coastal Reserve.

Still another example is Shackleford Bank, North Carolina, a barrier island southwest of Cape Hatteras that was once the site of Diamond City, a town of five hundred people who supported themselves largely by fishing. Buffeted by frequent storms, they lived on the island's sound side, in the protection of its large dunes, though a few relatively affluent residents also maintained informal oceanfront "camps."[2] Repeated storm damage, culminating in what must have been hurricanes in 1896 and again in 1899, convinced residents it was time to abandon their homes. Many of them jacked up their wooden cottages, rolled them on logs to the marshes on the island's shore side, and loaded them on barges for the short trip to the mainland. Survivors of this move, marked by small plaques commemorating their journey, can be seen today along the streets of Morehead City, the coastal town in Shackleford's lee. All that remains of Diamond City are the few tombstones in its tiny cemetery, shadowed by the scrubby ocean pines and visited only by the island's wild ponies and snakes.

More often, people simply kept out of the ocean's way in the first place. In the earliest European settlement of North America, there was practically no building at the coastline. The early settlers kept their houses on high ground, away from the heavy weather and the thin soil of the seashore. On Cape Cod, for example, settlers dismissed the oceanfront as the "back beach" or "back side" of the Cape, locutions that survive today among oldtimers. Until well into the twentieth century there were few settlements and hardly any buildings near the Cape's Atlantic beach — except the shacks maintained for the shelter of mariners unlucky enough to wreck their boats on the offshore shoals.

Henry Beston, the naturalist and author, wintered on this beach in a modest cottage he had built in the 1920s, and his account of that hard season, *The Outermost House* is still much read by beach-lovers. There were no roads along the beach then, he wrote, and "my knapsack remained the only ever-ready wagon of the dunes."[3] (Beston called his house "the Fo'castle," and it stood at the shore in Eastham until 1978, when it was destroyed in one of the worst winter storms ever in the Northeast. Everyone who knew him, and his love for the wild beauty of the beach, said he would have liked that.)

The same building patterns prevailed on the Outer Banks, where the settlers built their houses on the sound side, away from the erosion and storms of the ocean. The first hotel on the banks, a two-hundred-guest facility built in Nags Head in 1838, faced Albemarle Sound. It was not until the 1920s that people began to build cottages by the ocean, and even then they built movable cabins they could haul out of harm's way.

A century ago, it was easy to abandon coastal land. It was cheap, development was sparse, and the houses were modest enough to move.

Most land along the shore was regarded as useless except perhaps for grazing. Even in relatively urban areas, there was still room to move inland. In April 1888, for example, the Brighton Beach Hotel on Coney Island, a multistory frame structure with scores of rooms, was moved 450 feet inland when it was threatened by erosion. Its owner, the Brooklyn, Flatbush, and Coney Island Railroad, jacked the thing up and hauled it inland in one piece, using six locomotives, 112 flatcars, and twenty-four specially laid tracks. The structure moved "at a fast walk," *Scientific American* reported in its issue of April 14, 1888, adding: "No difficulty of any kind was encountered."[4]

But in the years since World War II, and particularly in the last twenty-five years or so, there has been a land rush at the coast. Demand for property there is intense and, overall, coastal real estate is among the most valuable in the nation. Landowners reap the largest return from their investment in this property by developing it to the limit and beyond.

As coastal land grows in value, beach houses are becoming more and more elaborate. The small dune-sheltered cottage of fifty years ago is a thing of the past. Today new-built beach houses have four, five, or six bedrooms, each with its own bath, and are equipped with with every sort of luxury. Some of these new houses are permanent residences, or second homes. But many are rental properties, which must be lavishly equipped if they are to command the high rents their amortization requires. Vacationers who once came to the beach to enjoy sea breezes now demand air-conditioning and cable television.

By 1990, there were few coastal communities in the United States with enough open land to allow owners of beachfront buildings easy access to another lot further inland. Besides, the value of many threatened structures lies in their nearness to the water. Move these buildings away from the perilous shoreline and they would lose, not gain, value.

So, rather than retreat from the beach, Americans are digging in.

As early as 1962, when the Ash Wednesday storm destroyed thousands of buildings along the East Coast, coastal scientists began saying that much of the coast was unsuitable for development and that efforts must be made to limit building there. On the whole, their efforts have been in vain. To be sure, there have been attempts from time to time to bring a little common sense to the frenzy of coastal development. Most of them have been at the national level, with federal regulations as stick and access to the federal treasury as carrot. The three major efforts are the Coastal Zone Management Act of 1972, the National Flood Insurance Program, enacted in 1968, and the Coastal Barrier Resources Act of 1982. Their record has been spotty at best. Sometimes they seem to have actually encouraged unwise development of the coast.

COASTAL ZONE MANAGEMENT ACT

Though it did not come first, the Coastal Zone Management Act was the law that formally put the United States government in the coastal policy business. Its goals were "to preserve, protect, develop, and, where possible, to restore or enhance the resources of the Nation's coastal zone" and to encourage states to develop and implement management plans "to achieve wise use of the land and water resources of the coastal zone."

The law required the participating states to draw up management plans, and by 1995, twenty-nine of the thirty-five eligible states and territories, representing 95 percent of the nation's coast, had adopted their own programs to conform to the law. First, they defined their "coastal zone," coastal waters and adjacent shorelands. Then, they identified "permissible land and water uses" and "areas of planning concern." These were troublesome questions, especially in localities where sentiment against government controls on land use was strong. Finally, each state had to set up some procedure, agency, or mechanism to exercise some sort of control over coastal land-use decisions.

Some states have acted vigorously under the act. North Carolina has banned coastal armor. Massachusetts and others have greatly restricted it. Still others, such as Florida and South Carolina, have attempted to limit building, by requiring that structures be set back an established distance from the beach. Though not then covered by the act, Texas enacted a law, the "Open Beaches Act," which theoretically allowed the state to remove any structure that, through the effects of beach erosion, ends up seaward of the line of vegetation.

Despite loudly voiced fears that the act was yet another attempt to quash local development with federal regulation, it would be hard to make the case that the Coastal Zone Management Act has stolen decision-making from coastal managers in individual states. Because beaches vary so widely from place to place, definitions in management laws are difficult to apply nationally or even within a given state. California, for example, has more than eleven hundred miles of coastline, as long as the coast from Boston to Charleston and encompassing almost as many changing landscapes. Also, the law's simultaneous demands for advancing development and encouraging preservation are often at war with each other.

In the end, most state officials have little appetite for battle with the powerful economic interests of the coast. After Hurricane Hugo, South Carolina caved in to the pleas of property owners and modified its law to allow some rebuilding seaward of the setback line. In Texas, the act was modified in practice to allow people whose houses were less than 50 percent destroyed to rebuild on the beach. And even the definition of "destroyed" is not clear-cut.[5]

NATIONAL FLOOD INSURANCE PROGRAM

The flood insurance program, administered by the Federal Emergency Management Agency, was established in 1968 with the ostensible goal of reducing the need for federal disaster relief by discouraging new building on the coast and in river floodplains. In return for adopting strict new construction standards (the stick), communities would be eligible for federally subsidized insurance (the carrot). Today the program covers tens of thousands of properties, insured for billions, in coastal and riverine areas around the country.

At first, the program was widely applauded on the coast, and with good reason. Correctly carried out, an insurance program is the best kind of planning for disaster relief. Requiring property owners to buy insurance shifts the burden of disaster spending onto them. And if they have insurance, the federal government does not have to step in with emergency assistance when the inevitable storm occurs. Besides, after the program had been in place for a while, it became clear that houses and other buildings constructed according to the program's specifications—especially its requirement that they be set on stilts or otherwise kept above the flood level of a one-hundred-year storm—suffered much less damage than those that did not conform.

And for some of the time, at least, the program's backers could brag that all this was carried out without taxpayer subsidy, because money taken in each year as premiums was more than covering the cost of paying claims. Hurricane Hugo, Hurricane Andrew, several bad winters, and a season or two of devastating Midwest floods put an end to that kind of talk. The program ran huge deficits, paid out of the federal treasury. These storms exposed the central, glaring flaw in the insurance program: it is betting the federal treasury *against* a sure thing: coastal flooding.

This problem was obvious at the outset, had anyone stopped to consider it: insurers who want to stay in business do not sell cheap policies to people who are sure to file substantial claims. That is why the program was needed in the first place. Until the advent of federal flood insurance, it was practically impossible to insure beachfront and nearby structures against flood damage, except through state insurance pools (where they existed) or underwriters such as Lloyds of London. This meant, among other things, that most aspiring beachfront property owners had to pay cash for their property, because most bankers would not issue mortgages for structures they could not insure. It also meant that most houses on the beach were modest affairs. If they washed away, so be it.

Now the federal government was insuring flood-prone beachfront properties for a few hundred dollars a year apiece. The availability of this insurance encouraged mortgage bankers to lend money for coastal development, fueling the coastal property boom. Advocates of the insurance program now say property values would collapse along the shore should the federal program be tightened. That would be "shorelining," they say, as unfair as the "redlining" that occurs when banks unfairly deny credit to people in low-income neighborhoods they outline with a metaphorical red line on their maps. Most social theorists—and even most bankers—denounce redlining as one reason for the persistence of social inequality in the United States. But it would be hard to find a logical reason to attack banks if they refused to finance mortgages for buildings in danger of washing away. As Steve Leatherman puts it, "Would you insure people who live on the lip of a volcano?"

Aware of the danger the program was in from risky coastal properties, FEMA proposed, in effect, that property owners be encouraged to remove their endangered buildings—either by relocating them inland or tearing them down. This plan, enacted by Congress as the Upton-Jones program, allowed owners of coastal property threatened with imminent destruction to claim 110 percent of its insured value if they demolished it. They would be paid 50 percent for removing it to a spot elsewhere. When this provision was enacted, many advocates of the program hoped it would encourage sensible retreat.

But the lure of the coast was too strong. By the mid-1990s, only a few hundred homeowners had taken advantage of the program, all but a few of them in North Carolina. Even after Hurricane Hugo, South Carolina produced only five claims under the program. Ultimately, it was abolished.

Frustrated, FEMA turned to the National Academy of Sciences, commissioning it to examine the issue. In 1990, an expert panel appointed by the Academy returned with recommendations: FEMA should consider erosion risk as well as flood risk in insuring coastal properties, and it should require communities that want to participate in the insurance program to establish setback lines to keep people from building in erosion hazard areas. Even the National Academy declined to take up the issue of how much the insurance should cost. The program had raised rates in the 1980s for new construction, and officials claimed that the program was "actuarially sound," but the figures did not take erosion risk into account. Still, the recommendations plunged the insurance program into an argument over whether it is possible to predict how soon a given stretch of beach would erode.

The recommendations would require costly additions to the extensive mapping effort the insurance program had already undertaken. It already had delineated areas where a one-hundred-year storm—the kind of storm that is so severe there is only a 1 percent chance one will occur in a given year—would leave enough water to produce three-foot waves (hydrostatic or "A" zones), and areas where waves would be more than three feet high, strong enough to challenge a building's foundations (hydrodynamic or "V" zones). Now the scientists wanted to add "E" or erosion hazard zones, where there was substantial risk of serious erosion within ten years. The scientists suggested that the insurance program deny coverage to new construction in this zone, but the mere existence of these maps, especially if sellers were required to disclose them to prospective buyers, would do a lot to discourage unwise building in erosion zones, they said.

Opponents of the change seized on the mapping requirements, saying it would never be possible for scientists to accurately predict when a given stretch of coast would erode. To a degree, they had a point—establishing erosion rates along the coast is complicated. It is even more complicated when the beach has been armored with seawalls, groins, or jetties, or even if it is near this kind of armor. Areas near inlets are still harder to deal with, as are renourished shorelines. To predict future erosion on a rebuilt beach, researchers must consider whether the projects in question meet minimum design criteria—criteria that, in some cases, are still being developed; whether relevant government authorities have agreed on how the structures or rebuilt beaches will be maintained; and whether the project has a reliable source of money to carry out these plans.

Nevertheless, scientists were sure they could produce adequate maps. Testifying on the proposal before Congress, Leatherman said he and his colleagues at the University of Maryland had learned how to digitize old maps and photos so that they could be analyzed by computer. Along most of the coast, he told a congressional hearing, aerial photographs go back sixty years or more, and in many places accurate maps go back two hundred years or more. The resulting accumulation of data, especially the abundant data of the last sixty years or so, is more than enough to make accurate predictions and provide community officials and citizen groups the information they need to make planning decisions in their towns.

FEMA adopted the scientists' recommendations on E-zones and setback requirements and submitted them to Congress as proposed alterations to the insurance program. Had they passed, the changes might have accomplished much of what needs to be done at the coast. At first, chances looked excellent. Among other things, an administrator of the insurance program, on what was literally his last day on the job, declared at a hearing that the only actuarily sound way to insure a building in a ten-year erosion zone would be to charge an annual premium equal to 10 percent of its insured value.

To be sure, the proposal had its critics among scientists. One of them was Bob Dean, the coastal engineer at the University of Florida, who called the bill "well-intentioned" but too vague and said it would not do enough to encourage sand-bypassing at inlets, something he believes would solve many erosion problems, especially on the East Coast. "If we are going to try to tell people they have to move back and abandon their property, let's first correct what we can," he urged.

But it was the real estate industry, officials in coastal towns, and beach-front property owners, many of them influential political contributors, who really turned the situation around. There was no need to alter the program — it had money in the bank, they said. And anyway erosion mapping would be expensive, difficult, and filled with mistakes. Their most telling argument, though, was this: if the federal government stopped offering subsidized flood insurance at the coast, property values there would plummet.

The proposed changes were defeated. The defeat left the federal government in a dilemma similar to the one it faced in the savings and loan crisis of the late 1980s. As the insurer of last resort, it pays the price when something goes wrong, but meanwhile it is state, not federal, authorities who are setting the rules.

Ironically, most people who own property in the coastal flood zones do not take advantage of the insurance. They may purchase it, if required to do so to obtain a mortgage, but in many cases there is no mechanism to ensure that they continue to pay the premiums and they drop it, apparently trusting in the federal government's long history of bailing out coastal areas after disastrous storms. Some in Congress are now suggesting that Americans living with any sort of natural disaster threat—be it tornadoes, earthquakes, or hurricanes—be required to purchase federal "disaster insurance."[6]

Meanwhile, though, private insurance companies are doing what the federal government cannot: they are withdrawing even further from the coastal insurance market. Ten small insurance companies went bankrupt after Hurricane Andrew, which produced $15.5 billion in covered damages, and profits of larger companies doing business in the state were severely affected. Companies began canceling policies or even paying coastal customers to take their business elsewhere.

Companies like Allstate and State Farm have limited their sales in coastal areas of Florida and North and South Carolina. In 1996, Nationwide Insurance Company, the nation's fifth largest insurer of homes, announced it was drastically reducing its sales of homeowner policies in coastal areas.[7] The state of Florida imposed a moratorium on policy cancellations and set up a reserve fund to reimburse insurers in case of catastrophe, but many policyholders have nevertheless seen a doubling in their premiums, deductibles, or both. Annual rate increases are in double digits. Eventually the state ended up establishing an underwriting agency that now ranks as the state's second largest home insurer.[8]

COASTAL BARRIER RESOURCES ACT

There is intense debate now as to whether the existence of the flood insurance program—and the safety net of federal disaster relief in general—create a kind of moral hazard that exists when people are free to engage in dangerous behavior, like building on the beach, because they know they are protected from its consequences.

The Coastal Barrier Resources Act (CBRA) of 1982 offers a kind of test of that theory. The act declared that "Coastal barriers contain resources of extraordinary scenic, scientific, recreational, natural, historic, archaeological, cultural and economic importance, which are being irretrievably damaged and lost due to development on, among, and adjacent to such barriers."[9] Federal assistance was contributing to this unwise development and should be halted. The act created a Coastal Barrier Resources System, based on an inventory, prepared by the Department of the Interior, of privately owned, undeveloped barriers. Development within the system is allowed, but the act prohibits most federal infrastructure support for barriers within the system, and bars spending on flood insurance, roads, bridges, sewer and water systems, and beach nourishment.

The act may have slowed the development of barrier islands it covers, but development has continued nonetheless. One of the places covered by the act and developed anyway is Topsail Island, an extremely fragile barrier island on the southern part of the North Carolina coast developed without federal infrastructure support. The island is subject to the wild swings of a migrating inlet, and much of it is eroding rapidly. Piles of bulldozed sand typically line the one roadway along the island. Pilkey called it "the worst development in North Carolina." In 1996, he was proved correct.

In July of that year, Hurricane Bertha came ashore nearby, chewing Topsail's already fragile dunes away and flooding its houses and condos. Less than two months later, another hurricane struck, and this one was far stronger. Hurricane Fran came ashore at Topsail at 8 P.M. on a Thursday night, September 5, with winds of 115 miles per hour. Its dunes gone, Topsail was defenseless. By the time Fran moved inland, two-thirds of its oceanfront houses were dumped in the surf, floating in the inland waterway, lying in the road in ruins—or simply gone. The island's water, sewer, and electric power infrastructure had been wrecked.[10] The town hall of North Topsail Beach was gone and its police station, housed in a trailer, had floated away.

In theory, because virtually all of this building was covered by the coastal barrier act, property owners should have been ineligible for any federal disaster assistance. In fact, federal and other aid flowed in for more than a year as homeowners rebuilt and workers removed ton after ton of debris from the fragile marshes behind the island.

Despite the assertions that the flood insurance program does nothing to encourage development, developers continue their efforts to expand its coverage area by chipping away at the CBRA boundaries. In the closing days of the 104th Congress, for example, they succeeded in removing some land on Florida barrier islands from the CBRA-designated system. They called the move "a technical correction" to legislation that had included the land by mistake. Opponents said it was just another example of opening land to development at (ultimately) taxpayers' expense.

Barring the federal flood insurance program from insuring buildings in the E-zone is not in itself an order to abandon the coast, but if it discouraged people from building or rebuilding there it would initiate a kind of rolling retreat from the beach. As sea levels rose, any kind of setback line based on natural phenomena—distance from natural dunes, vegetation line, mean high water, or the like—would theoretically roll inland with every storm.

In real life, however, coastal property owners would push right back. After the destruction of the Ash Wednesday storm, Robert Dolan of the University of Virginia was appalled to see the way people rushed to rebuild on the coast. "Someone once said, 'beach sands retain few scars to remind people of their instability,'" he wrote twenty-five years after the storm. "Rebuilding and new development soon resumed on the barrier islands. Wherever that owner of a damaged or destroyed home decided a location was too risky, someone was eager to take over the site."[11]

Since the Ash Wednesday storm, in fact, Americans have been moving to the coast in ever greater numbers. Almost half of all construction in the U.S. in the 1970s and 1980s took place in coastal areas,[12] and demographers estimate that by the year 2000, 80 percent of Americans will live within an hour's drive of a Pacific, Atlantic, Gulf, or Great Lakes coast.[13] By 2010, the National Oceanographic and Atmospheric Administration estimates that almost half the U.S. population will live in counties actually on the coast.[14]

Storms offer overbuilt coastal towns the opportunity to return to a sane level of development, but only if local officials have the will and the clout to stand up against the enormous pressure to rebuild. In town after town, this has not been the case. In fact, nowadays property owners do not merely rebuild. They replace storm-damaged buildings with even bigger structures. For example, when Hurricane Frederick struck the barrier islands on the coast of Alabama near Gulf Shores in 1979, it wiped out many of the oceanfront cottages there. By then, gulf-front property was far too valuable to waste on modest cottages, so the destroyed houses were replaced by high-rise condos that line the beach at Gulf Shores. The two-lane road that once served as the island's major thoroughfare is now a four-lane highway, built with federal funds, as were new water and sewer lines.

If regulations look stern enough to actually prevent rebuilding, they are challenged and undermined. South Carolina's setback requirements did not survive a storm of litigation threats after Hurricane Hugo. State officials who liked the requirements realized that the cost of litigating every challenge to them would break the state's budget, even if it ultimately won. Florida's "control line" is under constant assault. New Jersey restricted development within five hundred feet of the beach, but later modified its regulations to exempt structures containing fewer than twenty-four units, a gigantic loophole. In fact, New Jersey's record at the coast is terrible. The state, whose magnificent beaches have been chewed to pieces by development, established a "Shore Protection Management Plan" in 1981, but as its name suggests its emphasis was on coastal stabilization, largely through engineering.

In 1996, scientists at Rutgers, led by Norbert Psuty, reviewed this plan and produced their own recommendations. This time, the emphasis was in improving public safety and reducing economic losses from coastal storms. Their report was called "Coastal Hazard Management Plan for New Jersey." Instead of concentrating on preserving development in hazardous coastal areas, the report denounced it. "There is a justifiable concern that the coast is overdeveloped and that the millions of visitors to the New Jersey shore are exhausting the remnants of natural character and quality that once was so prevalent," it said. Meanwhile, the report went on, on 82 percent of the state's 127-mile shoreline erosion is classified as a "critical" erosion area and another 9 percent is suffering noncritical erosion. Only 9 percent of the shoreline is classified as stable.

New Jersey's coastal communities have drawn a line of development in the sand, the report concluded, but they can no longer afford to defend it. As a result, "planning efforts should focus on mitigation of the risk to public safety and not on the defense of property." Beach replenishment efforts should not take place in high erosion areas, it said, and when storms damage infrastructure it should be rebuilt inland.

Reaction to the report was swift and nasty. On the coast and in the state legislature, it was denounced as "a suicide strategy." The state Department of Environmental Protection scrapped scheduled public hearings on the report, and it remains to be seen if any of its recommendations will be adopted.

Another approach that creates a sort of rolling retreat is to forbid the kind of coastal armor necessary to maintain structures at the shore. Several states have done so, but they are having trouble making regulations stick. It may be difficult to reject a flooded homeowner's request to rebuild, but it is even harder to turn down a resident who only needs to throw a few boulders onto the beach to save a beloved home.

Perhaps because its coastal landscape is so vulnerable, one of the first efforts to ban hard stabilization occurred in North Carolina, which adopted a Coastal Areas Management Act in 1974. The act was under challenge from the outset. Initially, advocates of regulation hoped authority under the act would be centered in a professional staff of coastal planners. "This proved unacceptable to the affected local governments," according to David Owens, a professor at the University of North Carolina who participated in the creation of the regulations. "It was also opposed by private property owners and the development community, both of whom felt they would have more influence if the decisions were made locally by political rather than technical personnel."[15] A citizen commission was created instead.

Still, after loud and intense lobbying by people like Orrin Pilkey, the commission issued permit and setback requirements and, in 1985, banned hard stabilization of the beach. Though many state residents support the ban as the only effective way to maintain public access to healthy beaches, people are using one provision of the law to get around its intent.

That provision has to do with sandbags, which the law allows as a temporary measure to protect threatened structures until beach nourishment projects can be carried out or the buildings can be removed. But that is not how the regulation works in practice. People now look at sandbags as long-term protection, and the definition of sandbag has proved elastic; beach strollers in Nags Head or other communities on the Outer Banks often encounter houses, motels, or condominiums, their dunes eroded away, standing on the beach surrounded with walls of sandbags— and not just small canvas sacks, either. More and more, property owners turn to giant tubes ten or twenty feet long and four or five feet in diameter, covered in heavy "geotextile" fabric coated with an impermeable plastic that makes them waterproof. Once in place, the unsightly sandbags look and function like walls. Like walls, they eventually fail. But they are hardly "temporary."

The first major commercial challenge to the law came from Shell Island, a $22 million condominium resort built in 1985—after passage of the state's antihardening law—at Mason Inlet, in Wrightsville Beach. Like many inlets, Mason Inlet is highly unstable. Studies indicate it moved fourteen feet toward the condo building within ten years of its construction. The owners association of the 169-unit development sought permission to install various kinds of "inlet migration barriers," including a plastic wall eighteen feet high, much of which would be buried, and a two-hundred-foot length of interlocking steel sheets.

Lawyers for the owners of the condos, which are run as hotel suites by a management company, promised to appeal but said the appeal probably could not be heard in time to save the condos. Less than two hundred feet of sand remains between the condo and the ocean, and it is unlikely an appeal could be heard in time to allow construction of the barrier, even if it were ultimately allowed. Meanwhile, many owners were selling their units at a loss—to buyers willing to bet tens of thousands of dollars that the government would eventually step in and bail them out.

They were encouraged in this gamble by an earlier decision by the commission, which granted permission for construction of sandbag groins at nearby Bald Head Island. Largely uninhabited as late as the 1960s, Bald Head Island is a recent luxury development with acres of marsh and maritime forest, tennis courts, a golf course, and scores of condos and beach houses. But it lies at the mouth of the Cape Fear River—in fact, it forms Cape Fear—and its shoreline, like all cape shorelines, moves wildly. From the mid-nineteenth century until development began in earnest, sand had been piling up regularly there. In 1974, though, everything turned around. The sand that people had counted on began to erode away. "The actual trigger is uncertain," according to Kevin R. Bodge, a coastal engineer who has studied the island. Perhaps it was due to dredging changes in a nearby navigation channel, or the breaching of an offshore shoal in a storm. In any event, erosion on the southwest end of Bald Head Island has been severe. Homes that had been built with reasonable setbacks and prudent construction standards were suddenly in danger, or already lost. Heeding the pleas of property owners, the state coastal council allowed "an experiment" to test the use of the bags as sand traps. When sand was pumped onto the beaches later, the bags would hold it in place.

But in fact the sandbags used were geotextile tubes three hundred feet long and nine feet high, which looked and functioned exactly like groins along the 1.3-mile stretch where they were installed. Opponents charged that property owners had used their political influence and wealth to get the decision they wanted. Approval of the project, its opponents said, symbolized "the power of a wealthy elite to use political influence to divert coastal management policies to its own narrow interest."[16]

In any event, the experiment failed. The pumped sand washed away, there were cries that the sandbag groins were worsening erosion else-where, and within a few years the sandbags themselves began to fall apart, adding an unsightly mess to the island's troubles.

The development has already assessed extra taxes on waterfront homes and on vacation rentals to raise money to maintain the beach. And it is not clear what they should spend it on. Geologists believe that Cape Fear is in the process of migrating to roughly the position it had in the mid-nine-teenth century, and their prognosis for much of the resort is bleak. "Bald Head Island is going to have problems from here on out," said William Cleary, a coastal geologist at the University of North Carolina at Wilm-ington. "They've got themselves locked into a situation and they can't get themselves out. It's like an open pit and they're going to have to pour more and more money into it."[7]

It is an assessment that applies to many areas on the coast.

Many coastal property owners persist in asserting that any regulation that limits their right to build in erosion hazard areas amounts to an unconsti-tutional confiscation of their property. They have taken to the courts to ad-vance this view.

This argument was asserted most forcefully in a case involving David H. Lucas, an entrepreneur who in 1986 bought two oceanfront lots in a de-velopment called Wild Dunes, on Isle of Palms, north of Charleston, South Carolina. Records show he paid $975,000 for the two, which were separated by one lot between them. The intervening lot, as well as the lots on either side of his, all had elaborate new beach houses on them, and the island as a whole was high, dry, and newly developed with large, luxurious houses. Lucas planned to build similar beach houses on his lots and an-ticipated he would be able to sell them at a substantial profit.

But the appearance of his property was deceiving. Wild Dunes is near the north end of Isle of Palms, where Dewees Inlet separates it from De-wees Island to the north. Over the long term, coastal geologists believe, Isle of Palms is accreting slightly; that is, the island's shoreline is advanc-ing as it gains sand. But it is advancing in gulps, so to speak, and every pe-riod of accretion is followed by a period of substantial retreat.

These gyrations occur because the accretion is being fed by the ebb tidal shoals that form at the mouth of Dewees Inlet. When they are large enough, they begin to move south toward Isle of Palms, where much of the sand eventually moves onshore, causing the beach to grow. Meanwhile, though, a new shoal is forming at the inlet's mouth, trapping sand again. Then the beach at Isle of Palms begins to erode.

State planners were studying old maps and other records of these gyrations when Mr. Lucas was buying his lots, and the planners took them into account two years later, in 1988, when the South Carolina legislature passed the Beachfront Management Act. Citing the importance of beaches to the state's tourism economy, the act sought to protect the state's beach and dune system from damage by overdevelopment by banning the construction of permanent structures too close to the ocean. It drew the line for development based on forty years' worth of data about erosion and erosion rates.

To establish the line, they digitized aerial photos dating since the 1940s that showed how the beach had eroded and accreted in cycles over the years. (The planners disregarded photos taken immediately after hurricanes or major storms, which they felt would distort the data.) They used the photos to draw setback lines, which they placed forty feet landward of the farthest encroachment the sea had made over that time period.

Since Lucas's lots had been entirely under water as recently as 1963, and inundated twice a day at high tide as recently as 1973, the setback line was set well inland of his property. Though Lucas's neighbors escaped this restriction because they had already built on their lots, the South Carolina Coastal Council, which administered the act, declared that Lucas would be allowed to build no permanent structures on his property, only something like a small deck or walkway.

Lucas sued the council, asserting that its restrictions reduced the value of his property so much that they amounted to a condemnation or "taking." As a result, he argued, the state must compensate him for the value he had lost. He won the first round, when the trial court awarded him $1.2 million in damages. But the state appeals court reversed that decision, saying the state had a right to prohibit nuisances in environmentally sensitive areas. Lucas took his case to the United States Supreme Court.

Meanwhile, Hurricane Hugo had slammed into South Carolina. Threatened with a flood of litigation over the act's restrictions, the legislature altered it, substantially easing the process by which property owners could obtain variances allowing them to build. Nevertheless, Lucas pressed his suit, and the court, in a complicated opinion dealing with nuisance laws and the limits of the authority of state legislatures, ultimately ruled for Lucas.

In the end, the state of South Carolina bought the lots from Lucas, for about $1.5 million. While a few South Carolinians joked sourly that the state could use them to establish recreational facilities for children from low-income neighborhoods of Charleston, the state did the only sensible thing under the circumstances. It sold the lots for development, and a large house has been built on one of them.

A few years later, just as the coastal geologists had predicted, the cycle of accretion turned into one of erosion at the north end of Isle of Palms. By 1996, high tides were lapping at the foundations of homes on the lot and neighboring parcels. Trucks and earth-moving equipment, hired by the desperate homeowners, have become regular visitors to the beach, bulldozing sand into berms in front of houses on the threatened stretch of beach.

"The berm-building exercise is reminiscent of the never-ending task faced by fiddler crabs who must clean the sediment from their burrows after each high tide," says Stanley R. Riggs, a professor of geology at East Carolina University who has done extensive research on coastal processes. "The crabs, however, have evolved through the eons with these coastal rhythms and are in equilibrium with the changing high-energy processes. Immobile houses with their absolute deeds are very much out of equilibrium."[18]

When the Court issued its ruling in the Lucas case, many environmentalists predicted it would be a disaster for the coast. That has not yet proved to be the case. For one thing, the Court distinguished between regulations that eliminate all value and those that diminish it and declared that mere diminution of value is not a taking that necessarily requires compensation. Second, the Court hinted that the regulation might have stood if the legislature had based it not on economic arguments but rather on the needs to preserve lives and property from needless hazard. In general, legal scholars say, when the state takes property to confer a benefit (as South Carolina proposed to do in aiming its regulations at improving economic development) the state must compensate the property owner. When it takes property to prevent a harm, there is no right to compensation. The South Carolina legislature made a mistake drafting its beachfront regulations by referring to their benefits for tourism and economic development. As David J. Brower, a lawyer and planning expert at the University of North Carolina, put it, "those are not a reason to regulate land use."

He and others cite other steps state and local authorities can take to make regulation stick:

- Ensure that all economic value is not eliminated. This is not always as easy as it seems, as Brower showed by inventing a hypothetical situation in which he goes into the building business. "Suppose I buy land for $100,000. Its best use is one hundred units. Now a regulation says I can build only fifty. Then twenty-five, then ten, then two. Then a platform for a tent with only 120 square feet, provided I haul out all my waste and use only bottled water. Where is the line? This case doesn't say." Still, he added, "you would rally have to work at it to render a piece of land absolutely valueless," and the Supreme Court has long established that mere diminution of value, however substantial, is not itself a taking.
- Identify viable uses of the land that conform to the regulations and allow the owners of land to do as much as possible with them. ("Be fair," Brower advises.)
- Consider administration remedies, such as converting building rights on one piece of property into additional building rights on another. Courts have upheld this kind of density transfer.
- Review real property law and nuisance law to make sure the owner is not already barred from undertaking the proposed development.
- Consider the state's public trust interest in the land in question. "Be ready to use it creatively," Brower said. "I don't mean to do anything untoward. But there are many ways it can be used creatively."

As Professor John C. Keene of the University of Pennsylvania law school noted at the National Hurricane Conference in 1995, if there is "an essential nexus"—a relationship in nature and extent—between the legitimate public interest and the permit requirements, regulations are more likely to stand up in court.

Regulators must also consider questions they will be asked if they are liberal with building permits on the coast:

- What costs does building there impose on the rest of the taxpayers?
- If you get a permit and build and your building is destroyed, can you sue because a public agency gave you a permit to build in an obviously unsafe place?
- If I build in a V-zone and my house is destroyed and its debris damages my neighbors' homes, do my neighbor have a cause of action against me? (A good question, but it will be moot until better ways are devised to determine whose debris has damaged whose house.)
- Some legal theorists (and not a few antigovernment activists) propose compensating property owners for any loss of value occurring through

regulation, a suggestion that raises a host of complicated questions, quite apart from issues of how it would be financed. To cite just one, does the payment of compensation for lost value give the state an interest in the property? If so, the owner's cost basis would theoretically be reduced, and the state might even be viewed as a tenant in common with the owner, with a voice in other decisions about the property, such as whether it should be sold. Also, if the property is mortgaged, who should receive the government compensation, the lender or the owner?

All this theorizing is interesting, but it ignores several important points.

First, people should not demand compensation in exchange for obeying the law. Laws may bar any number of lucrative land uses—marijuana cultivation, gambling, operating a factory in a residential zone—but landowners do not ordinarily expect compensation, even if zoning laws are changed after they buy their property. As Professor John A. Humbach of Pace University School of Law put it in a presentation at the National Hurricane Conference in Atlantic City in 1995:

> The fact is that governmental actions constantly affect property values without the least thought that taxpayer dollars should compensate the owners affected. The Federal Reserve raises interest rates and bond traders lose value. Congress deregulates the airlines, and several of them go broke. It deregulates the savings and loans with financial results that are even worse. By definition, laws that prohibit socially deleterious uses of property also deprive owners of whatever profit there was in making the antisocial use. Yet, the constitution permits such laws, including most laws regulating land use, because otherwise people could freely engage in conduct that is too socially unacceptable to allow.[19]

Second, who pays the price if someone like Lucas builds on his lots and the structures are later destroyed through the action of erosion and storm waves? As things stand, the answer is: everyone. The lots are eligible for federal flood insurance.

Third, why is the property valuable in the first place? Often, it is valuable because the government is subsidizing it in some way. In the Lucas case the subsidies are clear—ordinary infrastructure support, but also flood insurance, disaster relief, etc. The government is not obligated to give people this kind of construction help, especially in an erosion hazard zone.

Fourth, if natural resources are to be preserved, laws and regulations must adapt to advances in our knowledge about the natural world. Swamps, wild animals, obsolete buildings were once regarded as obstacles to progress. Today, wetlands, biological diversity, and historic structures are regarded as blessings to be preserved.

Finally, if it is called a "taking" if government regulations deprive owners of some use of their property, then it must also be a taking if a private citizen acts to diminish or destroy a natural resource that belongs to the public at large. It is a truism that rational individual decisions, actions that are privately optimal, can produce undesirable outcomes for society as a whole. This phenomenon is obvious at the shore. Rational people making rational economic decisions have crowded the coast to the point that its value as a resource, for them and for society at large, is reduced. And instead of compensating society for this loss (never mind the disaster-related costs they exact), the users are continuing to profit from the degradation of the landscape. Researchers who study the value of natural ecosystems call this situation—the conflict between narrow short-term incentives and long-term environmental health—"a social trap."[20]

How to get out of this trap? The answer is far from clear. It is difficult to confront individual resource-users with the consequences of their actions, especially when the effects are not precisely known or have not yet occurred, or when links between cause and effect are indirect, conditions that often apply at the coast. One approach might be to require builders to pay some sort of fee or post a bond or otherwise provide a source of funds to repair environmental damage caused by coastal development. Is such an arrangement politically feasible? It is impossible to say.

When Republicans won control of House of Representatives in 1994, they promised severe restrictions on the kind of regulatory takings Lucas and his allies inveighed against in his suit. Environmentalists in and out of the White House beat them back. Though property owners are eager to free themselves from the regulatory yoke, it may be that Americans overall value the protections regulations provide. Meanwhile, legal battles over the fate of the beach seem certain to continue. Some legal scholars estimate that two-thirds of the most recent Supreme Court decisions on takings involve coastal property.

Problems with rule-making, local disagreements, and the extreme cost of defending the developed shoreline are leading more and more people to think the answer lies in turning the entire problem over to the federal government. Ironically, though, this idea is being heard just as federal officials are realizing that maintaining coastal development will demand an endless flood of money. They want the federal government to do less, not more, at the beach. In 1996, for example, the Clinton administration proposed taking the federal government almost entirely out of the beach nourishment business. Eroding beaches are largely a local problem, the theory was, and they should be handled locally.

The idea was little more than a whisper in the ear of Corps of Engineers budget-writers, but it echoed loudly on the coast. Almost immediately, coastal communities, business executives, and organizations, led by a Washington, D.C. lobbyist, formed the American Coastal Coalition to lobby for continued support of what they call "shore protection." If the White House was going to stop paying to rebuild eroded beaches, they would approach Congress instead.

The Corps of Engineers, the nation's leading beach-builder, supported their effort. James R. Houston, an ardent advocate of beach nourishment, began making its case in scientific journals, conferences, and elsewhere that beach nourishment is a highly effective technique for maintaining beaches. And beaches, he declared, are potent revenue-generators. Travel and tourism is one of the nation's largest and fastest-growing industries, he pointed out again and again. Other countries, notably Spain, Japan, and Germany, spend far more than the United States does on beach protection and replenishment, he said.[21] Even some opponents of coastal development added their voices, largely because they feared that a sudden disappearance of federal beach nourishment dollars would lead desperate oceanside communities to turn to even worse tactics, such as seawalls.

At the urging of the coalition, coastal state lawmakers in both the House and Senate formed coastal caucuses. The new coalition drafted legislation and, to its delight, lawmakers introduced measures in both House and Senate. The identical bills, "the Shore Protection Act of 1996," made it an official mission of the Corps of Engineers "to promote shore protection projects and related research that encourage the protection, restoration and enhancement of sandy beaches, including beach restoration and periodic beach nourishment, on a comprehensive and coordinated basis by the Federal Government, States, localities and private enterprises." The corps would be required to carry out projects authorized and funded by Congress and—perhaps even more important—contribute to maintaining the projects for their lifetime, typically fifty years.

The proposed legislation declared that "a network of healthy and nourished beaches is essential to the economy, competitiveness in world tourism, and safety of coastal communities of the United States." The effort succeeded—the central ideas of the legislation were included in a water resources bill passed with little notice in the waning days of the 104th Congress.

One way out of the mess on the coast would be to require people who live or invest there to bear the risks and cost themselves. "Let 'em take their licks," is how Leatherman once put it. But even he backtracked immediately. Though he prefers the parks on Assateague Island, with their four hundred campsites, he recognizes that others prefer the bright lights of a place like Maryland's playground, Ocean City. "It's a place the masses enjoy," he said. "Ocean City can swell to four hundred thousand people on a weekend. I would say you can't let Ocean City fall." Even so, he went on, people who think coastal development can be preserved indefinitely are living in a dream. "It's no secret any more what's happening on the coast," he said. "It might have been a few decades ago. We didn't have good data on erosion rates. Now we do, and we know about 90 percent of the coast is eroding."

At a conference on Long Island, Robert Dean of the University of Florida offered this scenario: "I would say, 'Okay, you're in a very hazardous area. If you build here and if you sell, you have to inform the people buying the property of the history of the area. If the structure becomes seaward of the mean high water line and stays there for a year, we're going to ask you to move your house.' " Property owners would get a terrible jolt if they heard news like this. The market value of some coastal property would surely fall. But eventually it would stabilize at a more realistic level. "People will adjust," Robert Sheets says. "They're not forced to adjust today."

In hopes of learning more about their coasts' behavior, a few states such as New Jersey, California, Florida, and South Carolina have established programs for regular monitoring of shorelines and shoreline erosion, usually through aerial photos and by profile surveys taken regularly at designated benchmarks. New Jersey conducts profile surveys every fall at approximately one hundred sites, along 114 miles of coastline. Florida has established more than thirty-five hundred survey sites, identified by fixed concrete benchmarks, one thousand feet apart. Each site is visited every

three to five years, oftener in the event of severe storms. In a tacit acknowledgment of the fragility of its beaches, the state has installed a second set of markers, five hundred feet inland of the dune line, to insure that
measurements can continue into the future.

Recently, the Federal Emergency Management Agency has been
studying the use of the satellite technology of the Global Positioning System to map shoreline change. In trials in Mississippi and Delaware and at
Duck, antennae were set up at control sites and then linked to instruments
in four-wheel-drive vehicles that drove along the beach at the high tide
line. Preliminary reports suggest this survey method may be substantially
better than traditional aerial photography.[22]

One problem is the vast divide between the scientists whose findings
should inform coastal policy and the people who must make policy and
carry it out. As a group, researchers are unwilling to enter the messy regulatory arena, especially on an issue like coastal development, where tem-
pers can literally flare into violence. (Pilkey once spoke at a meeting on
seawalls at which participants were assured that police would be present
to maintain calm.) Scientists want to do research, not policy. Some of
them actually look down on policy makers as failed academics, calling
regulation "the revenge of the C student."

"Funding favors narrowly directed research," Douglas Inman said at
the Woods Hole retreat, adding that he advises fellow scientists not to
get involved with public policy because it counts against them in tenure decisions.

Besides, many scientists are not temperamentally suited to drawing up
policy guidelines. Managers want yes or no answers. Scientists don't think
that way. Theirs is a world of probabilities and error bars in which anything less than 95 percent sure is not sure enough. So it was hardly surprising that when the NRC convened its experts at Woods Hole and they all
agreed that development was moving too far and too fast on the coast, the
recommendations they came up with had little to do with the practical decisions facing officials in coastal towns. Most of the recommendations
were for more money for coastal research.

Meanwhile, many activist groups like the fact that environmental regulation is fragmented, because it makes it easier for them to tie any development proposal in knots for years. If there were uniform regulations, uniformly administered, that would not be so easy. Even developers who
despise regulation will say they would prefer firm regulations over the
loose, variable system in place in many regions today.

The Southeast Light on Block Island, a small island in Rhode Island Sound, has little of the romance of the Cape Hatteras Lighthouse. Still, Rhode Islanders who have been to the island love the somewhat portly red brick structure and, like their compatriots to the south, were alarmed when erosion of the island's Mohegan Bluffs threatened to send their beloved lighthouse tumbling into the ocean. So in 1993 it was hoisted off its foundations and hauled to safety 245 feet inland.

When the 50-foot lighthouse was built in 1873, it was the tallest and brightest in New England, and there were 300 feet of land between it and the edge of the bluff. By 1992, when the decision was made to move it, it had only 60 feet of clearance, offering only a 20-foot safety margin of the stable ground the movers needed to run their equipment around the building without risk of falling down the 150-foot cliffs. And 5 feet of that was gone by the next year, when the work began.

There had been some talk at the Southeast Lighthouse Foundation, which had bought the lighthouse from the Coast Guard, of attempting to stabilize the cliff by intensive planting. But the idea was given up as impractical, in part because the cliff would continue to erode at its base and threaten the entire bluff with collapse. The group also discussed armoring the base of the cliff with stone but abandoned the idea because, even if it worked, it would only worsen erosion elsewhere on the island.

Instead, they decided to move the lighthouse back from the edge of the cliff. The $2.6 million move got underway on August 11, 1993, even as the movers hoped the existing bluff would survive long enough for them to complete the project. The move itself ended on August 27, as Hurricane Emily approached Rhode Island.

All in all, the move went smoothly. The movers, the International Chimney Corporation of Buffalo (also tapped for the Cape Hatteras move) and Expert House Movers of Sharptown, Maryland, ran beams diagonally through the basement walls; ran stabilizing bars through the windows of the house that formed the lighthouse base, to lower its center of gravity; and reinforced its tower. Without the tower reinforcement, "it might not have fallen apart but we definitely would have had a problem with it shifting," said Merle Copeland of International Chimney, who supervised his company's work. The whole thing was rolled ever so slowly along tracks, first to the north, then to the west, then north again, covering about a foot per minute. The movers wanted to move the two-thousand-ton structure on paths parallel to its walls, fearing a diagonal track would risk a collapse.

"The only thing that caused us any concern was when we excavated under the tower foundation," Copeland said. "It turned out to be nothing but a pile of rubble. The architectural drawings showed it was built out of step cut granite blocks." Perhaps, he speculated, there had been corruption in government building contracts even then.

For Copeland, who hails from Buffalo and whose previous job had been a smokestack project in the South Bronx, the isolation and small-town atmosphere of Block Island was something of a shock. "It took me a while to get used to the island," he said. "Nobody locks their car. Nobody locks their house." But he found the months of work rewarding. "We moved it back 245 feet, so the estimate is it's safe for between eighty to a hundred years," he said. "It depends on how fast the bluffs erode. It would be a shame to see it go over the edge."

CHAPTER TEN

For Sale

They paved Paradise,
Put up a parking lot.
—Joni Mitchell,
Big Yellow Taxi

Martha's Vineyard, an island off Cape Cod, is famous for its beaches, especially the twenty-mile stretch of barrier spits and grassy dunes that runs along its south coast from Chappaquiddick in the east to the cliffs of Gay Head in the west. Much of this beach is private, or open only to residents of one island town or another. But one part of the beach, in the town of Edgartown, though privately owned, has been used by everyone on the island for decades—since the days before islanders thought much about who owned the shoreline.

Vineyarders call it South Beach. Generations of island sweethearts courted there; later, they took their families there to picnic. Parents taught their children to fish at South Beach; in the fall, fathers and sons hunted waterfowl flying across to the ponds and salt marshes behind it. Spring after spring, islanders trapped herring making their annual spawning run into a creek that empties at the beach. And always it was a place to enjoy the sun, the wind, the flight of the gull. For many, South Beach summed up what it meant to live on the Martha's Vineyard.

So islanders were stunned to read in the *Vineyard Gazette* on July 10, 1987, that thirteen acres at the heart of South Beach had been sold to a developer, the Beach Cove Realty Trust, which planned to close the beach to the public—to everyone, that is, except 250 people willing to pay $25,000 each for access rights. After August 1, the developers said, the price would rise to $55,000.[1]

The developers explained their plan in a full-page advertisement in a later issue, describing the offering of deeded access rights to thirteen acres of South Beach as "a new and exciting concept." But deeded access rights are not unusual on Martha's Vineyard or elsewhere in Massachusetts and Maine (originally part of the Massachusetts Bay Colony), the two states where private property on the beach extends all the way to the low water mark. In these states, people can be barred from walking on the beach at all, even below the high tide line on the "wet beach," which elsewhere is part of the public's domain.

Martha's Vineyard has several beach "associations" controlling access to particular stretches of coast. At least one was started in the early 1930s to help the members of the family that owned the beach pay inheritance taxes on it. In the old days, access to these beaches might be through a small gate, its key left under a nearby rock for all to use. Now, though, the fences are more substantial, every association member has a key—and keys to Martha's Vineyard beaches sell for as much as $100,000.[2]

As a result, beaches open to the public are all the more precious. South Beach was the island's principal public beach, and now Vineyarders would have to make an appointment with a real estate agent to venture onto it. And few island families could afford the $25,000 introductory price.

Within days, the islanders had mobilized to fight the sale. In an editorial July 14, the *Gazette* declared, "The message of what historians will call plunder at South Beach is this: The Island is for sale and nothing is safe or sacred. Not even South Beach, this most cherished public symbol of what the Vineyard is all about." The editorial concluded with a call to arms: "It is time for the Vineyard to wake up and act. There is a war going on here."[3]

Vineyarders took the message to heart. One Edgartown resident set up a card table on a Main Street corner with a petition urging the state legislature to initiate a bond issue to finance the purchase of coastal property. Within hours, eight hundred people had signed it. Another took a similar petition to the beach itself and vowed it would take a bulldozer to drive him off. A big Boston law firm offered the town its legal firepower. Beach-lovers organized themselves into Concerned Citizens for South Beach; people from all over the country, most of them summer vacationers on the island, sent donations to support it.[4] Members of the Edgartown Board of Selectmen (its governing body) and the chief of the town's police force warned developers they would never be able to keep trespassers off the beach.

The truth was, though, that until developers threatened to snatch it away from them, islanders had not thought much about who owned the beach. It was privately owned, but except for a brief period in World War II, when the navy took it over, South Beach had always been open to the public. From 1957 to 1977, by agreement with the owners, the town of Edgartown had maintained and supervised the beach, and since then had paid those costs plus a nominal sum to lease the beach for the summer. Eventually, everyone assumed, the state would buy it. But no one had felt any need to rush the purchase, especially since, as is often the case on the Vineyard, legal title to the property was clouded by ancient questions about lot lines, inheritances, and unpaid taxes.

By the end of July, the letters, petitions, phone calls, and public protests were having an effect on state officials, who finally began moving to take the property by eminent domain. At the same time, the Massachusetts secretary of state began looking into whether the sale of shares in the beach would violate the state's securities exchange laws. And the town had won two important court victories: purchasers of the access rights would be notified that their purchase was the subject of litigation, and the new owners would be barred, at least temporarily, from keeping people off the beach. Meanwhile, the town was preparing legal action questioning the sale in the first place, saying that although the town had paid to lease the beach, it had actually acquired the land through a kind of adverse possession, in that "the town and the public's use of the beach has been open, continuous and notorious."[5]

Two months later, toward the end of September, the developers approached Edgartown officials with an offer: if the town would drop its legal action, they would give the town about seven acres of beachfront. Town officials rejected the offer. "As far as I am concerned, this is not a dealable commodity," a town park commissioner declared in an interview with the *Gazette*.[6] "It is a public resource and we should have it."

The storm ended early the next month, when the Massachusetts legislature began action on legislation for a bond issue providing more than $500 million for purchasing open space, including more than $40 million for coastal property. Three weeks later, the Department of Environmental Management announced it intended to buy as much as 77 acres of South Beach and would overcome title problems somehow. By the next spring, as the town prepared once again to set up its life guard stands on South Beach, the state was on its way to keeping South Beach public indefinitely.

State officials said they had learned a valuable lesson, just in time. "When you see New Jersey, Delaware, Maryland, and the Florida coast, you realize we're not too late," said Karst Hoogeboom, the landscape architect for the state Department of Environmental Management, which was charged with drawing up plans for the beach. "But too late is just around the corner. Someone has got to say, 'This is valuable. This is precious. This is our heritage, and we might have to give up a little bit in order to preserve it.'"[7]

James Gutensohn, commissioner of the state Department of Environmental Management, said: "Before the bond issue was passed and the developers were trying to condominiumize the beach, I said in many ways South Beach and its protection is a test for people as to whether or not government works. It is a test of the ability of government to implement what people's hopes and dreams and visions are. One of those dreams is the protection of special natural resources, places where people can enjoy themselves, places of scenic and aesthetic beauty."

There is only one sure way to preserve the coastal landscape from unwise and unsafe development: buy it. But for a long time, rising coastal real estate values made the idea so outlandish that people would not even consider it. ("Buy the coast!?" Katherine Stone, the sand rights advocate, exclaimed when asked about the idea. "With what money? We don't have any money to fix the roads or help the homeless.") Costs have not dropped any lower. But, perhaps in desperation over continuing blight on the beach, more and more people are beginning to fight money with money, simply buying up coastal property to keep it away from developers. Sometimes government agencies buy the land, as on Martha's Vineyard. Sometimes private individuals donate it to a government or private conservation agency, or take it out of development through conservation easements. Sometimes small contributions to environmental groups eventually add up to enough cash for a coastal purchase. Sometimes local jurisdictions establish taxing mechanisms to provide funds to buy land.

People who are skeptical about this approach should look to Britain, where the National Trust has acquired and preserved sizable stretches of coast in England, Wales, and Northern Island. The trust, established in 1895, received its first gift of coastal property in 1897 after developers announced they would build a large hotel in Tintagel, on the coast of Cornwall. The announcement prompted a group of coast-lovers to purchase a neighboring headland, Barras Nose, and put it in the trust's control. Nowadays, the trust accumulates money so that it is able to move quickly when a likely property comes on the market. Since 1965, when a fund-raising project called "Enterprise Neptune" was launched by the Duke of Edinburgh, the trust has raised enough money to purchase more than 450 miles of undeveloped coastline. By the mid-1980s, it owned the white cliffs of Dover, one-third of the entire coast of Devon and Cornwall, as well as large stretches of coast in Dorset, Norfolk, Yorkshire, and elsewhere. It hopes to buy at least another 400 miles.

Groups such as the trust promote their work with musical talk about preserving the treasures of nature for future generations, and indeed this goal is a noble one. But there is another reason to consider buying the coast to preserve it from development, and more and more people are beginning to realize it: tourism, on which so much of the coastal economy depends, depends in its turn on the natural features that draw vacationers in the first place. Otherwise, the tragedy of the commons will be enacted again and again, as private landowners exploit the coast until it is wrecked. It is ironic that ordinary people or taxpayer-financed agencies should have to raise money to protect coastal businesses from themselves, but it is evident that those businesses cannot or will not do it for themselves.

Though few Vineyarders made the point aloud, plenty realized that tourists come to the Vineyard in large part to enjoy its beaches, particularly South Beach. Without it, the island's shops, restaurants, and hotels would suffer.

But as the songwriter Jimmy Buffet once put it, "Paradise ain't cheap." Since the end of World War II, the price of coastal land has risen more or less steadily, and in the 1980s and 1990s it skyrocketed. Even as officials in Massachusetts were vowing to protect South Beach, they were wondering how they would find the money to pay for it. Funds that once would have been ample were now barely a downpayment. "Ten years ago, $25 million would have seemed like an absolutely enormous amount of money," Gutensohn, the environmental management commissioner, said during the South Beach affair. "It is still a large amount, but land prices are obviously going very high throughout Massachusetts and particularly along the ocean."[8]

Perhaps because it has the most money, the federal government has done more beach-buying than anyone else. Its National Seashore program today owns miles of beachfront property, most of it on the Atlantic and Gulf coasts. In almost every case, land has been added to the system only over the intense objections of local development interests.

The seashore program has its roots in the 1930s, when a group of people on the Outer Banks of North Carolina decided a "Coastal Park for North Carolina and the Nation" might mitigate both the economic woes of the Depression and the eternal erosion of their beaches. Their campaign culminated in 1937 when Congress passed legislation authorizing the creation of the Cape Hatteras National Seashore. The authorizing legislation allowed the federal government to acquire seashore property but only through land donation, or through funds donated to buy land. Still, it was a start.

The first donations—nearly one thousand acres—came from wealthy businessmen, among the many who had bought large tracts of land for private hunting clubs. Land acquisition did not begin in earnest until after World War II, and by then many residents of the Banks were having second thoughts. They had welcomed the prospect of jobs the park offered during the Depression, but now oil companies were nosing around the Banks, buying up mineral rights and planning test wells. David Stick, a historian of the Banks whose father, Frank Stick, was one of its earliest and most ardent preservationists, wrote: "People who had enthusiastically supported the Cape Hatteras Seashore park program in time of depression when it meant jobs and income now as bitterly opposed it in the face of the prospect of oil royalties, and they even succeeded in securing passage of a North Carolina law prohibiting further donation of Banks land for park purposes."[9]

The building boomlet that followed the oil companies led many to believe the park idea would die, but in 1952 foundations created by the children of the industrialist Andrew W. Mellon gave the state $618,000 to purchase land for the park, on the condition that the state match the donation. The state legislature appropriated the matching funds four days later. Finally, the park was assured.

Wayne Gray, the retired coastguardsman and lifelong resident of the Banks, remembers the anger of those days. Even though settled villages (Rodanthe, Waves, Salvo, Avon, Buxton, Frisco, Hatteras, and Ocracoke) were excluded from the park, he said, their residents hated it. "The park came in and took our land and told us what we could do. 'You can hunt here and you can't hunt there.' People didn't have money to fight it in court. That was it."

The development of the last few decades has changed their minds. Many people now realize that their ocean beach is the magnet that draws visitors to the Banks. Besides, they have only to look to Kitty Hawk, Kill Devil Hills, and Nags Head, north of the park, to see the fate they escaped. In those towns, hotels, condos, souvenir stands, and all the rest crowd the narrow barrier to the point that even the five-lane asphalt swath down the middle cannot handle the traffic. "People realize how the park has saved the banks," Gray said. "If the park service hadn't come in it would be developed from Oregon Inlet to Hatteras Inlet. The island couldn't stand it. I am very thankful the park service came in. They took land from my family, but I don't care."

Once a seashore has been authorized by Congress, the creation of the park itself begins with the establishment of its boundaries. Then the government begins acquiring the land within, by purchase, condemnation through eminent domain, or donation. Existing development may be exempted from the park, as was the case on Cape Hatteras, or homeowners may be allowed to remain for the rest of their lives, or for fifty years, or some other limited time period, as is the case at the Cape Cod National Seashore in Massachusetts. At the Fire Island National Seashore on Long Island, preexisting villages are not subject to condemnation as long as they maintain zoning restrictions approved by the secretary of the interior.[10]

The park service divides each national seashore into zones, depending on how the land will be used. In natural zones, there are few facilities other than picnic tables here and there, and people do not disturb the landscape much. In historic zones—the area around the Cape Hatteras Lighthouse, for example—the park service conducts activities related to preservation and education. These areas typically have heavy use. The same is true of the development zones such as visitor centers, campgrounds, concession areas, and beaches protected by lifeguards. Finally, there are zones for park development, primarily infrastructure to support the public's use of the park, and special use zones for Coast Guard stations, radio towers, and the like.

Today, the parks are:

- the Cape Cod National Seashore, which includes the Massachusetts coastline along 39 miles from the elbow of the Cape at Chatham to the curving spits at Provincetown. It was established in 1961, the first seashore park authorized with access to federal funds for purchasing land.

- the Fire Island National Seashore, twenty miles of beach on the south shore of Long Island, New York, plus seven islands in Great South Bay.
- Assateague Island National Seashore on a 37-mile-long barrier island on Maryland's eastern shore.
- Cape Hatteras National Seashore on a 75-mile stretch of beach on the Outer Banks of North Carolina.
- Cape Lookout National Seashore, including several islands south of Cape Hatteras.
- Cumberland Island National Seashore off the coast of Georgia.
- Canaveral National Seashore, where Florida waterfowl coexist with NASA's more exotic birds.
- Gulf Islands National Seashore, which runs along the Gulf Coast from Florida to Mississippi.
- Padre Island National Seashore in Texas.
- Point Reyes National Seashore north of San Francisco, California, the only national seashore on the West Coast.

Most of these parks are heavily used. The Fire Island National Seashore, established in 1964, is practically within commuting distance of New York City. One third of the nation's population is within a day's drive of the Cape Cod National Seashore. Despite heavy use, the park service has managed to preserve the wildness of much of the parks' landscape. This preservation was not always easily won. New York's notorious "master builder," Parks Commissioner Robert Moses, fought for decades to build a four-lane highway down the middle of Fire Island to "anchor" it. Real estate interests backed the idea, but it sank under the opposition of conservationists and island residents. At Assateague, park officials planned a six-hundred-acre concession area with a motel, saltwater pool, fishing pier, and shopping area. This plan died in the 1970s when the park service acknowledged that erosion was a fact of life on barrier islands and that, therefore, it was inappropriate to build extensive permanent facilities on them.

Some parks evolved despite the opposition of powerful economic forces. Cumberland Island National Seashore took its present form only after public opposition to a proposed development and another donation from a Mellon foundation, which allowed its purchase. Otherwise, the developer Charles Fraser might well have succeeded in his plan to create another resort like the one he built at Hilton Head, South Carolina.[11] The seashore at Point Reyes was established through an unusual business collaboration. Owners of the land, which had been used for ranching and farming, wanted to log it and then build luxury resorts. Proponents of a park battled to a compromise: the government would buy the land, but people who lived in the park would be allowed to remain for fifty years. And twenty-six thousand acres inside the park would remain a ranching zone. The landowners got less than they wanted from this deal, but more than simple condemnation proceedings might have afforded them. And the park backers were happy. They thought ranching would be a picturesque addition to the park's rugged landscape and, on the whole, they were right.

One park, the Gulf Islands National Seashore, owes its existence as much to history buffs as nature-lovers. The impetus for its creation came from efforts to preserve a series of nineteenth-century coastal forts.[12]

In some ways, it seems remarkable that there are any beaches left for purchasers to preserve. Often, beaches remain pristine only because some catastrophe prevented their development, thereby allowing a government or private agency to step in.

Sometimes the catastrophe was financial, as in the creation of Island Beach State Park, one of the few stretches of the New Jersey coastline that bears any resemblance to what nature designed. Henry Phipps, Andrew Carnegie's partner in the steel business, bought the land in 1926, intending to turn it into a resort. The stock market crash of 1929 stalled his plans, and he died the next year. The Depression and then World War II meant more delay, and eventually the state bought the property from his estate. Today it is the only long stretch of coast that does not require constant, expensive replenishment, and it is enjoyed by hundreds of thousands of people each year.

Financial problems also resulted in the purchase by Currituck County, North Carolina, of what was to have been the site of a large development on the northern stretch of the Outer Banks, in the town of Corolla. The site was the old Whalehead Club, built in the early 1920s by Edward Knight, a wealthy businessman whose wife, an avid hunter, was barred from the men-only hunting clubs that flourished on the banks before World War II. The developers were caught in the financial squeeze of the late 1980s and ended up selling the property, which the county purchased using funds from the occupancy tax levied on rental cottages. Today, the elaborate clubhouse, an eccentric mix of beaux arts and arts and crafts styles, is headquarters of the Currituck Wildlife Museum, which exhibits hunting memorabilia and a large collection of carved decoys, and the Whalehead Preservation Trust, which is helping to finance clubhouse preservation.

Disaster of another sort prevented development on Assateague. The federal government had been eyeing the island for park purposes since the 1930s, but bills introduced in Congress to finance its purchase had died in committee.[13] The agency now known as the Fish and Wildlife Service established a National Wildlife Refuge on the Virginia portion of the island in 1943,[14] and developers gave the state of Maryland more than five hundred acres in the 1950s, but developers purchased the northern end of the island and laid out a network of streets. They even paved some of them, including a wide, fifteen-mile asphalt road down the middle of the island, which they dubbed Baltimore Boulevard.

The Ash Wednesday storm of 1962 wrecked much of their work and gave potential buyers second thoughts. Government authorities bought the land, and in 1965 it was named a national seashore. Today, all that remains of the developers' grand plans are a few slabs of asphalt; a road sign marking BALTIMORE BLVD.; and a natural history trail, including an exhibit explaining the futility of trying to build on the coast.

Padre Island was similarly headed toward development when the Depression and a 1933 hurricane stopped development in its tracks. By the time the economy revived, Texans had already established the Padre Island Association, whose goal was the creation of a state park there. They, in turn, were delayed by World War II and litigation over ownership of the land. But as development pressure resurged in the 1950s, political support for a park grew as well. Padre Island National Seashore was authorized in 1962.[15]

A model for private land conservation, and the oldest private land trust in the world, is the Trustees of Reservations, established in 1891 to "conserve the Massachusetts landscape." Its founder was Charles Eliot of Boston, a partner of Frederick Law Olmstead, the nineteenth-century landscape architect who designed many of America's finest parks. Eliot proposed the creation of an organization of volunteers to acquire and maintain parcels of land "which possess uncommon beauty and more than usual refreshing power . . . just as the Public Library holds books and the Art Museum pictures—for the use and enjoyment of the public." Eliot died in 1897, but his ideas have spread far and wide. The National Trust in Britain, the Nature Conservancy in the United States, and a host of other organizations are modeled on the Trustees of Reservations.

In Massachusetts, the Trustees owns and maintains scores of endangered areas or historic sites, many of them on or near the ocean. The organization encourages donations of money and land, often by what it calls "creative" means. In 1995, for example, the Trustees financed the purchase of a 56-acre parcel on the West Branch of the Westport River, near Rhode Island Sound, by carving out a handful of house lots for sale on condition the new owners agree to conservation easements.

Taking its cue from the Trustees, the Nature Conservancy, established in 1950, has also become a major buyer of coastal property. Unlike the Trustees, whose interest in landscape preservation has led it to acquire ownership or management rights for almost twenty thousand acres of beach, marsh, riverbank, meadows, woodlands and historic gardens and homes, the Conservancy's goal is to preserve endangered species. In large part, it works by buying and preserving critical habitats. Unlike the Trustees, the Conservancy usually turns around and sells the land to government agencies or others who agree to keep it out of developers' hands. For example, the Conservancy spent $900,000 in 1995 to acquire South Williman Island, a 2.765-acre island in South Carolina, and then sold it to the state for inclusion in the Ashepoo, Combahee, and Edisto Basin National Estuarine Research Reserve. The state paid the Conservancy enough to cover the purchase price and its legal costs. (Earlier, the state had paid the Conservancy for four other islands the Conservancy had acquired through purchase or donation.)

This process—a sale and then a second sale—seems cumbersome, but it offers two advantages. In part because of the tax laws, the Conservancy can often negotiate a lower price than the land would otherwise have commanded on the open market. Also, the Conservancy can move much faster than any government bureaucracy. By staying liquid, it can purchase land the moment it becomes available. Then it can hold on to the land until a state legislature appropriates funds or voters approve a conservation bond issue—by which time the parcel might otherwise be under development.

The Conservancy also encourages landowners to protect their own land by donating or voluntarily establishing permanent restrictions on development. Such "conservation easements" can be a boon to land-rich but cash-poor people who want to hold on to their land but who cannot afford the high property taxes levied on "developable" real estate. Once development is barred on the land, its market value and its tax bill drop.

Sometimes, the Conservancy acts by stealth, as it did in purchasing barrier islands on the Virginia coast. In 1969, a New York investment group announced plans to build a retirement and recreation complex on three of the pristine islands there, Smith, Myrtle, and Ship Shoal. The builders planned houses, malls, hotels, boat basins — plus a nine-acre nature reserve — near a convention center. All of this development would be linked together and to the mainland by bridges and causeways through the island marshes.

Financial trouble, feasibility questions, and a possible water shortage on the islands made the developers reconsider. With a large donation from the Mary Flagler Cary Charitable Trust, the Conservancy bought the islands and later expanded its purchases to include barrier islands running sixty miles down the Virginia coast. Today the properties are known as the Virginia Coastal Reserve.

Many nearby residents resented the Conservancy's interference with what they hoped would be lucrative development. Landowners began to resist Conservancy purchase offers, seeking offers from more development-minded buyers. Owners of one island finally agreed to sell to a corporation called "Offshore Islands Inc." Later, it became known that the corporation had been set up by the Conservancy solely to make the purchase. Meanwhile, an organization called the Allegheny Duck Club was buying up farms on the mainland across from the barrier island, particularly land with deepwater frontage, attractive to developers. Its leaders turned out to have ties to the Conservancy.

Only a private organization such as the Conservancy can engage in the kind of wheeling and dealing that made the reserve possible. Today, it is the largest landowner on Virginia's Eastern Shore and its holdings are a priceless natural resource for plants and a vast variety of shore birds. Indeed, the National Science Foundation has designated the area a Long-Term Ecological Research Site. Its tidal marshes are such prolific nurseries of fish that they have been called "protein factories." Though the islands lack even such basic amenities as restrooms and drinking water, they are also visited by thousands of beach-lovers every year.

Nevertheless, the Nature Conservancy recognizes the importance of jobs and a vibrant economy. It was a major supporter of the decision to move the threatened Block Island Lighthouse, in part because of the lighthouse's importance as a tourist attraction on the island. "What we are attempting to do is to preserve the ecosystem, preserve a way of life, and promote sustainable economic activity," said Keith Lang, director of the Conservancy's Rhode Island chapter. This means agriculture, fishing, and tourism, he said.

Together with the Conservancy, Block Islanders have sheltered 20 percent of their 6,400-acre island through purchase or easements. They have succeeded even as the island's tourism economy, moribund for most of the twentieth century, slowly revives. Block Island, dubbed "the Bermuda of the North" at the turn of the twentieth century, has a year-round population of only eight hundred, but may have thousands of visitors on a summer weekend.

The Conservancy has used its broad repertoire of real estate techniques to make all this preservation possible. It obtained one twenty-nine-acre parcel at an advantageous price by offering its owner a tax deduction and a ranch in Montana in exchange. In another instance, it served as an intermediary when the state purchased land from another twenty-six owners, helping its owners sell the land in installments to reduce their taxes.[16]

The efforts are backed by a host of other organizations, including the Block Island Conservancy, the Audubon Society, the Block Island Land Trust, the Rhode Island Department of Environmental Management, and the Champlin Foundation, which since 1986 has given $2 million a year for land preservation in the state.[17] But most people on the island say the biggest factor in its preservation success is the willingness of its landowners to sacrifice their own potential profits for the good of the island. "There are people coming to us," Lang said, citing the owner of six building lots who put a house on one and offered conservation easements on the others. "They recognize that it's too nice a place to play a role in its death."

With these groups as examples, land trusts have sprung up all over the coast—indeed, all over the country. Using tax incentives, conservation easements, and other techniques, they stretch conservation dollars by encouraging the owners of coastal land to sell it at a discount or take it out of development through conservation easements, reaping potentially large tax benefits in either case.

In some cases, individual citizens have banded together to save a stretch of coast, usually when they realize how much their own enjoyment of life—and property values—depend on preservation of open beach. For example, residents of Santa Barbara moved fast when a 69-acre parcel they knew as "the Wilcox property," the last piece of undeveloped coastline in their town, came on the market. The property, on a bluff whose woods and meadows lead to sandy cliffs that tumble down to a narrow beach, had been used by members of the public for years. Many have been married there, and a few people have even had their ashes scattered from the bluff toward the sea.

Small Wilderness Area Preserves, a group comprising mainly residents of homes near the property, had tried for years to engage the property's owners in a discussion about buying the place. But without some sort of sales agreement, they could not raise the money, and until they had some money, the owners did not want to negotiate.[18]

Finally, the group won a commitment from Santa Barbara County for $1 million toward buying the land, and in mid-January, 1996, the owners put the property on the market. They asked $3.5 million and gave the group until the end of February to come up with the money. In a frenzy of activity, residents sold baked goods, held benefit poetry readings and musicales, and even cooked dinners and charged friends to attend. Children set up an apple-juice stand (a local market contributed the juice) and donated the proceeds to the cause. Thousands of donations, 80 percent under $100, poured in as the local newspaper, the *Santa Barbara News-Press*, ran daily progress reports. By February 29 the group was still $600,000 short, but at the last minute, an anonymous donor offered to make up the difference. The purchase itself will be a complicated transaction involving the owners of the property, their bank, and the Trust for Public Land, a San Francisco–based group that will formally purchase the land and turn it over to a government agency in Santa Barbara.

Many coastal areas have established official, government-sanctioned agencies to buy land for preservation, supported by fees on land purchases or development.

Voters on Martha's Vineyard approved the Martha's Vineyard Land Bank in 1986, when the island was undergoing intense development. "Farming declined, centuries-old pastures and fields were left to knot into vines and shrubs. The 'freedom to roam' was curtailed as fences were erected across trails, beaches were gated off and hunting was restricted," the land bank said in its annual report in 1996. Vineyarders realized that their island's character was irrevocably changing for the worse. The land bank was their response.

It imposes a 2 percent surcharge on most real estate transactions on the island and uses the money to buy land. In its first decade, the land bank purchased 1,100 acres of land—1.5 percent of the island—much of it on or near the island's coastal ponds and bays. The land bank allows public access to all of its property. (It also allows hunting on some of its properties, and access to them is limited in season.) Some areas are used for farming and others are reserved for wildlife. In all cases, trails avoid sensitive areas, and parking is limited to small clearings at trailheads. "Balance is key in land bank property management," the 1996 report declared in 1996.

Depending on how much buying and selling is going on, the Martha's Vineyard Land Bank may accrue $2 million, $3 million, or more in a year for future purchases.[19] But money remains a problem. "Revenue . . . is modest compared to need," the report said. Fortunately, its conservation efforts are assisted by other land trusts like the Nature Conservancy and the Sheriff's Meadow Foundation, another private trust that concentrates on preserving wildlife habitat.

Other states, counties, and localities around the country have established similar land-buying agencies with dedicated funding sources. But often these efforts face intense opposition from real estate and other development interests. Despite the immediate success of the Martha's Vineyard Land Bank, and a similar operation on the nearby island of Nantucket, efforts to establish a land bank on Cape Cod, begun in the 1980s, met defeat after defeat. When another attempt was begun in 1998, many on the Cape viewed it as their last chance.

"We've talked about this for 20 to 30 years," Eric T. Turkington, a state representative from Falmouth, told the *Boston Globe*. "We talked about how the traffic was going to get to the point where it's insufferable and about how development would get to the point where our water supply is threatened. Well, we're there now. the only tool out there that's available to us is buying land. The window is closing on us."[20]

But even when they have money to spend, they do not always find it easy to obtain the land they want. One of the most contentious problems is how to establish the value of the land in question.

In the case of South Beach, the *Gazette* suggested in an editorial that the sales offer was a ploy to raise the price the town and the state would eventually have to pay for the property.[21] If 250 shares were sold, even at the introductory price of $25,000, the value of the property would no longer be the $2.65 million the developers had paid for it. By then, cynics calculated, the individual purchases would have raised its value to more than $60 million. In the end, the price the state eventually paid for South Beach was far higher than it might have been, had developers not already begun selling their $25,000 shares.

Similar questions were raised when the state of North Carolina expressed interest in buying Bird Island, a small barrier that lies on North Carolina's border with South Carolina and is one of the state's last privately owned, uninhabited islands. As an island, it is nothing much. Most of its eleven hundred acres are underwater at high tide. In the entire place, only about thirty or forty acres can be developed, and their value is questionable, since the island can be reached only by boat. Still, it had all the virtues of an undeveloped coastal barrier: white sand, rolling dunes, a scrubby maritime forest, abundant bird life, and acres of glistening marsh. More than one visitor has described Bird Island as "the most beautiful place in the world."

State officials, many of whom had been vacationing nearby for years, had long talked of taking steps to enable the state to buy the island, whose value, assessors estimated, was just over $1 million. These calculations were thrown into disarray, though, when the owner of the island applied for permission to build a bridge over an inlet to the island. One-acre lots in the most beautiful place in the world, if you could get there by car, would surely bring as much as $500,000 each on the open market. The million-dollar island suddenly turned, in theory anyway, into a $20 million property.

The ensuing months brought bitter disputes over whether state law allowed a bridge to the island, since it would have to cross an inlet and traverse part of the marsh. Opponents said it was a clear-cut case: the state should not allow this kind of hazardous and environmentally unfriendly construction. To the owners, though, the case was equally clear-cut. There had been a bridge there decades ago, and now they were only trying to replace it.

In January 1996, the North Carolina Coastal Resources Commission finally voted to prohibit construction of bridges in inlet hazard areas, effectively barring the proposed causeway and bridge across Mad Inlet to Bird Island. Now advocates of purchase must raise the money, tapping the state's Natural Heritage Trust Fund, the North Carolina Coastal Land Trust, and other organizations. Frank Nesmith, the local resident and nature-lover who has been leading informal nature tours of the island for years, accumulated $7,000 in pledges in two days the following spring. It remains to be seen, though, who will set a value on the property—and how.

Advocates of preserving land from development must also overcome fears—legitimate or concocted—that taking part of the coast out of economic circulation costs jobs, reduces prospective tax revenue, and otherwise saps business vitality. Land has been preserved for the enjoyment of the public, this argument goes, but at a cost far larger than its purchase price.

Though there have been few thorough studies of what happens to a coastal town's property values and tax base when its beach is preserved from development, there is ample evidence that this argument is wrong. In the first place, preserving the beach from development obviates the need for any of the armor, replenishment, or other high-cost maintenance. Some economists who have studied the beach would even argue that a coastal town never reaps in tax revenue what it has to spend to protect houses built right on the beach.

But there is another advantage. Keeping builders off the beach may deprive a town of the high tax assessments of oceanfront property, but it also gives everyone else in town a greater share in the value of the beach—and, as a result, higher property values and maybe even a higher tax base as well.

Seaside, Florida, illustrates this effect. Seaside is a planned community established in the early 1980s on the panhandle coast west of Pensacola. The town is justly famous for the architectural restrictions and zoning requirements its designers enacted in their effort to re-create the small-town feeling of the old coastal South. These building rules have resulted in a tightly built network of streets lined with wood bungalow- or cottage-style houses with deep windows, shady porches, and picket fences. The community is built for walking, not driving. The houses are built on small lots and none is more than a quarter of a mile from the town's commercial center. Sandy paths, set off by picket fences, run behind every house.

The idea was to re-create the kind of close-knit community feeling that prevails in small towns, where everyone knows everyone else and people spend much of their time on foot, walking to shops, the post office, church, and the like, and greeting their neighbors on porches and across the back fence, rather than traveling from retail outlet to retail outlet hermetically sealed in automobiles. That is not quite the way it has worked out. Though several hundred houses have been built, only a couple of dozen, at most, are occupied year-round by full-time residents. Most of the rest belong to absentee owners who rent them out by the week or month to vacationers attracted to Seaside's ambience. (This ratio will change, Seaside officials say, if owners move to the town when they retire, or if telecommuting gains greater favor.)

Architects and urban planners are most interested in the design restrictions that force builders to construct houses with a limited vocabulary of architectural elements—deep windows, porches, peaked roofs, and picket fences—gathered by the town's designers, Andres Duany and Elizabeth Plater-Zyberk, in tours of the small towns of the Gulf. From a beachgoer's

perspective, though, the most interesting thing about Seaside is that relatively few of its buildings are directly on the beach. The main road
through town runs along open dunes interrupted only by a few two-story
houses (romantic places often rented to honeymooners) and a cluster of
shops and restaurants. The town itself is built on streets that run straight
back from the beach, crossed by curving avenues that spread out in widening half-circles as they move back from the ocean. Access to the beach is
provided at the end of each street by walkways over the beach dunes at the
end of the streets. Old-fashioned wooden gazebos stand at the apex of the
dunes, and residents can sit in their shade and enjoy the beach view before walking down for a swim.

Seen from the streets, the gazebos offer constant reassurance that the
beach belongs to everyone in town. As architects and city planners have
noted with approval, Seaside is a self-conscious embodiment of a new way
of considering public space in American society. In designing the town, its
planners turned the beach into a public realm that everyone in Seaside
has a share in, not a private realm enjoyed only by those with houses on
the water. Zoning requirements on the streets that run toward the beach
even require unusually large setbacks, to maximize the view of the dunes.
The code allows people to build as high as they like, but anything over two
stories tall must have have a "footprint" of no more than 215 square feet.
The requirement ensures that no one can forever block anyone else's view
of the water.

The move has had one dramatic result: property values in Seaside are
high. Though its designers hoped Seaside's mix of apartments, condos,
and houses with varying lot sizes would attract residents of many income
levels, the town is almost exclusively the domain of the upper-middle
class. And unlike most coastal towns, where houses on the beach rent or
sell for enormous amounts and houses even one or two rows back are
worth far less, values at Seaside are much more uniform. It is hard to escape the conclusion that the decision to keep houses off the beach has
boosted, not depressed, the town's property values.

The town's layout provided another, even more valuable benefit when
Hurricane Opal struck the panhandle coast of Florida in October 1995.
The storm, one of the most severe to strike the state in the twentieth century, left the region with extreme property damage—except in Seaside.
The storm cut into the dunes, but their height, and the large volume of
sand they contained, prevented the wave and storm-surge damage that left
other shoreline communities wrecked.

Other coastal towns have attempted to officially mark the beach as an amenity that belongs to everyone in town, but often their efforts succumb to the pressure for buildings right on the beach. For example, Carl Fisher, the original developer of Miami Beach, did not like the idea of building on the ocean beach. His original plan called for building Miami Beach hotels on the bay side of the barrier island and limiting the ocean side to public parks and relatively few estates for the rich.

Property he owned at the heart of what is now the city on the beach was restricted to private homes until after World War II, and until then people who drove up and down Collins Avenue, the central traffic artery, could actually see the ocean. By 1947, though, a change in city zoning regulations allowed building pools and "accessory buildings" seaward of a setback line established in the 1930s. The next year, the regulations were amended again to allow two-story buildings. Within a decade, high-rise hotels were marching shoulder to shoulder up the beach. Today one can drive for miles on Collins Avenue and see hardly a sign that ocean surf pounds behind the condos lining the road.

If people have enough luck, money, and savvy to save coastal land from development, the question then becomes: What should they do with it? How should they manage things so that people can enjoy the beach without ruining it?

The first step should be a coldhearted realization that anything that interferes with natural processes at the beach is likely to cause unexpected and potentially severe problems there or somewhere. Managers should work with the landscape and its natural processes, not against them. Decisions should be site specific—and site-specific recommendations depend on thorough knowledge of what the site is like and how its ecosystem functions. This task is difficult, in large part because beachgoers refuse to acknowledge the contradiction inherent in maintaining a pristine beach while making it accessible to people.

Beach managers must also decide who may use their land, and when. Some organizations interested in species protection more than recreation restrict public access. Others, like the Trustees, welcome people. For example, at their reservation on Chappaquiddick called Wasque, the extreme eastern end of Martha's Vineyard, the Trustees have constructed trails for four-wheel-drive vehicles and movable walkways to allow walkers easy access to the beach. After a storm, these walkways can be removed with a forklift and set down again on the altered beachscape.

Wasque itself was a triumph of the "buy the coast" philosophy. A scrubby sandplain, for hundreds of years it had been used for little more than grazing. But in 1913, it was subdivided into 775 quarter-acre house lots, which were offered for sale at $185 apiece. A few were sold but war, depression, and another war intervened, and it was not until the late 1940s that the few residents of Chappaquiddick began to fear their small island would be swamped with new building. Looking for someone to bail the island out, they approached officials of the town of Edgartown, which includes Chappaquiddick, and other government agencies, with no luck. Finally they turned to the Trustees, which stepped in and bought the entire place. Fishermen were among the major contributors to the purchase price.

The Trustees recognizes that these people and other beach users are an important constituency and source of support for its conservation efforts. So it tries to accommodate the public as much as possible. Though there is still plenty of conflict over access to the fragile landscape, these efforts dampen some of the grumbling when the Trustees limits four-wheel-drive traffic or closes beaches in the summer nesting season of endangered shorebirds.

Elsewhere, beach managers try to balance preservation and business interests through ecotourism. There will be less pressure for development, they reason, if business owners realize their revenue depends on the well-being of their beach. But this kind of balance is not so easy to achieve on a barrier island. Practically anything that draws visitors exacts a cost on the landscape.

Indeed, at island parks such as Assateague and Chincoteague, on the Delmarva Peninsula, the attraction causes the problem. These islands are the home of feral ponies—supposedly the descendants of horses from wrecked Spanish galleons but actually the offspring of animals left on the islands more recently to graze. The ponies have a voracious appetite for anything green, particularly the beach grass that is key to holding island sand in place. Wherever they go, they weaken island structure.

But tourists love them.

At Shackleford Bank in North Carolina, part of the Cape Lookout National Seashore, park workers were able to remove goats, pigs, and the descendants of other farm ruminants from the island but, by public demand, the horses were allowed to stay. Now researchers from Duke University and elsewhere are measuring exactly how much damage the horses are doing to the landscape. They have constructed "exclosures," fenced areas where the horses cannot graze. As they expected, marsh grass grows vigorously there and traps sediment. Where the horses are kept out, the island grows. As a result, many who want to preserve the beaches argue that the horses must be removed, one and all. Others take a different view. They argue that if the horses encourage people to visit the parks and learn about the fragile barrier island landscape—that is, if they help build a constituency for coastal preservation—the benefit they bring to the coast will far outweigh the damage they do.

The horse population is kept under control at Chincoteague by an annual roundup in which ponies are swum to the mainland and sold at auction. At Assateague, park officials are trying birth control, injected by dart. In developing areas, though, the horses may have met their match. In the village of Corolla, at the northernmost reach of the Outer Banks, horses that used to roam empty dunes found themselves wandering between luxurious beach houses or congregating in shopping center parking lots, where they found shady carports and freedom from flies. But purchasers of the luxury homes and shoppers at the arcades complained about the horses' foul smell and their ubiquitous droppings. As traffic increased, there were more accidents involving horses, fifteen of which were killed. In March 1995, the fourteen remaining horses in Corolla were captured and penned on the north end of town, near the Corolla Lighthouse. Their roaming days are over.[22]

Tourism represents about 6 percent of the national economy, and in coastal areas the figure rises to 20 percent or more. This fact is often cited as a rationale for continuing development. But it is just as forceful an argument on the other side.

Coastal communities are finally realizing that the well-being of their tourism businesses depends on a healthy coastal environment. More and more people understand that growth and jobs depend not on building on the beach but rather on keeping off it. After all, hotels, gambling parlors, T-shirt shops, and the like can be enjoyed anywhere. But a beach can only be enjoyed at a beach.

In Florida, where the health of the tourism economy is a life-or-death issue, the state legislature has initiated a ten-year program to raise $300 million per year to purchase land. Local government agencies are also starting to buy land. For example, residents of Brevard County, Florida, a coastal county immediately to the south of Cape Canaveral, established an Environmentally Endangered Lands program that, among other things, emphasizes biological diversity and species protection as it makes decisions on preserving coastal land.[23]

Though they will have to do without the tax revenue they might have had from coastal development, property values in a town with large expanses of public coastline may rise more than enough to compensate. It is routine for real estate brokers to advertise nearby nature preserves as valuable amenities that, in some cases, can greatly increase a property's value. Town officials will no longer face the costs associated with the inevitable storm damage and rehabilitation and protection of the beach. Government agencies pondering the high cost of buying coastal property should ponder as well the high cost of allowing it to be developed. Inevitably, they will confront infrastructure demands such as water and sewer lines, plus bridges, causeways, and other access facilities the new island residents will surely demand. If they factor in the costs associated with storm damage and disaster relief, officials may decide their only cost-effective approach is to solve the problem before it begins, by buying the land.

People who build on the oceanfront create infrastructure and service demands on the rest of the community that may exceed their contribution to its tax base. And almost any kind of development degrades the environment to at least some extent. The mere presence of a condominium changes the character of a beach, even if it never casts its shadow on the sand, spews sewage, or churns out bag after bag of trash.

In the end, when the last nail has been hammered into the last two-by-four of the last building on the last bit of developable land, the people in a beach town may realize that their economic growth is inextricably tied to the quality of their environment. But by then it may be too late.

Many times it is local people who are the biggest beneficiaries of development, and they may make the biggest push for it. If they own land, suddenly they find themselves surrounded by people waving checks, offering to pay what seem to be outlandish prices for acreage they have been barely making a living with as farmers. The local politicians, landowners, builders, lawyers, and real estate brokers will be right beside them. If they are not careful, though, they will lose more than they gain.

Wynne Dough, curator of the Outer Banks History Center in Manteo, North Carolina, grew up north of town and has been watching this process for years. He cites development on the northern reaches of the Banks, in and around the town of Corolla, as an example of how a town can trade a priceless landscape for large but transient profit. Until the mid-1980s, that stretch of the Banks was almost unpopulated, its rippling dunes empty except for the wax myrtle, waterbirds, and ponies. "There wasn't a paved road up there," he said. Today, oversized beach houses, each containing perhaps four thousand square feet or more, march over the dune. "Now, if you went up that way, you went past dwellings owned by Governor [Doug] Wilder of Virginia and Representative [Dick] Gephardt of Missouri and assorted Duponts and Phippses. There are $2 million spec houses. That gives you an idea of the optimism people have down here. They know eventually someone is going to come down here on a three-day weekend and write a check."

But the owners of these nouveau cottages do not do much for the local economy, he went on. "People stay in their own cottages. They come down with a trunkload of groceries and they leave a can full of trash behind and that's about the extent of their contribution to the local economy." Meanwhile, they cause extensive traffic congestion and need additional police protection, fire services, and medical care. "They place considerable demands on the infrastructure and they contribute very little," Dough said.

In the process, the character of the community can disintegrate. Nowhere has that process been more vivid than on the sea islands of South Carolina and Georgia. There, an entire culture is being wiped out as African-American communities, many of them remnants of the Gullah settlements where African traditions and language had lived for hundreds of years, are bought out wholesale to make way for "plantation" developments such as Hilton Head, one of the first planned coastal developments of the modern age. Compared to other developments, Hilton Head is remarkably sensitive to its environment, although its beach, built from sand dredged to drain marshy areas of the island, is artificial and in continual need of nourishment. (In 1991, Charles E. Fraser, developer of Sea Pines Plantation, the island's first resort complex, called for a "freeze" on building there.)

But more than beach has been lost. There is no sign of the island's original inhabitants—until they or their children drive over from their mainland homes for their jobs in the new resorts. A history of Hilton Head sold at many island gift shops tells in some detail of early French and Spanish exploration, pirate days, the Revolution, and the Civil War, but then it skips ahead one hundred years with hardly a mention of the people who lived on the island from then until it was developed as a resort.[24]

The Penn Center, an academic and cultural preservation organization headquartered in Beaufort, South Carolina, is attempting to preserve some of the vanishing history of South Carolina's coastal people. But as one commentator noted, many of them "have lost their history and died."[25] In fact, it is not uncommon nowadays for residents of the island to say that before it was developed it was "uninhabited."

As the quarterly journal of the South Carolina Sea Grant Consortium puts it:

> When Hilton Head Island began prospering in the 1960s, many observers hoped that [surrounding] Beaufort County would be the coast's model for the future. Indeed, the county is a success in some respects: it has the highest per capita income of any county in the state, and a high percentage of college-educated residents. But Beaufort County also has a large number of poor people and a sizable number of residents without high school diplomas.
>
> Many of the poor and undereducated residents work in resort service jobs. Resort service jobs . . . are usually low paying. . . . Service employees are not unionized or organized. During the winter off-season, lasting up to five months a year, many employees must survive on diminished work hours. And because workers often cannot afford expensive housing near their work, they must endure long commutes. . . .
>
> In some areas dominated by tourism and retirement, workers have few job opportunities outside the service economy. . . . [O]ne third of Beaufort County's residents are underemployed, earning less than $10,000 a year.
>
> Some Beaufort County residents cannot pay local taxes on their land—property that has been in their families since the Civil War. These taxes pay for better services demanded by the tourism and retirement industries. . . . In Beaufort County, dozens of locally-owned parcels of land have been sold in the past year due to non-payment of taxes.
>
> Other areas on the coast . . . are now drifting into Beaufort County's pattern of a large, mostly white upper class; a large, mostly black service class; and a diminishing middle class.[26]

Wayne Gray, who spent his childhood on the Banks and most of his adult life at the Coast Guard station at Oregon Inlet, was born in 1940 in a frame house without electricity in the village of Avon. In those days, before the construction of the paved road that now runs almost the full length of the banks, travel was by water, or by cart along the beach, and each isolated village had its own church, its own store, its own post office. Some people even said they could tell which village people came from by variations in their speech.

"When I grew up as a boy you knew everybody in the village," he said. "Avon was seven miles from Buxton, fifteen miles from Salvo. And you more or less stayed in your village. To go to Buxton was a big deal."

Gray, who is retired from the Coast Guard now, works as a law officer in the Dare County Sheriff's Department. Compared to the old days, when everyone knew everyone else and there was no crime to speak of, "today we've had so many people move in from the outside that you hardly know anybody," he said. There is far more crime than there used to be.

It would be naive to think that the lifestyles and economics of an earlier day can continue into the twenty-first century's global economy. (Few of us, for example, are prepared to live without electricity.) But it would be wise to ponder what we lose in our rush to develop the coast.

Lena Ritter is one who knows. A spare woman with graying hair, she has lived all her life in coastal Onslow County, where she owns thirty-seven acres of land she has no intention of developing. "I just want to live my quiet life," she says. But her life is hardly quiet. She comes from a fishing family, and nowadays she spends much of her time fighting pollution and other ills of overdevelopment. In 1950, she said, pollution had closed 5,000 acres of North Carolina's shellfishing grounds; by 1990, 315,000 acres were permanently closed.

Once developers start turning the coast into a vacation or retirement spot, she says, the characteristics that brought people there in the first place begin to change drastically. The new arrivals "get bored after a while. They want that nightlife that they left back in the city." There must be a new medical facility—unneeded perhaps, until now—"so developers can say they are within x minutes of medical care."

"You kids are going to have to buy back my childhood," she said recently to a group of college students Orrin Pilkey had brought on a field trip to the North Carolina coast. "I had a wonderful childhood out on the water, but it's going now. In twenty years it will be gone."

Let us give Nature a chance;
she knows her business better than we do.
—Montaigne, *Essays*

In mid-August 1995, a tropical storm in the northern Caribbean intensified and headed west. Moving at little more than a walk, it passed over Bermuda and toward the Outer Banks of North Carolina. Its winds were barely hurricane strength, just over 74 miles an hour, but their reach was great. Extending almost 180 miles from the storm's eye, they sent heavy surf crashing onto beaches from South Carolina to Massachusetts. By August 16, Hurricane Felix was wreaking havoc on town after town along the Atlantic coast.

In Sandbridge, Virginia, for example, heavy surf was pounding the sheet-pile bulkheads that were all that stood between many oceanfront houses and the surf. At high tide, storm waves broke over the bulkheads, many of which collapsed with the weight of the water behind them. Everywhere, the waves scoured at the bases of seawalls, causing many to settle and collapse. Residents returned after the storm to find windows blown out, roofing shingles blown away, and porches and decks washed out. Their beach, already narrowed to a few feet or less at high tide, was covered with the debris of collapsed bulkheads. Like residents of other built-up beach areas, they surveyed the destruction, added up their losses, and wondered why the weather seemed to be getting worse and worse.

Martha's Vineyard was battered also. The surf struck its south shore with such force the pounding could be heard more than a mile inland. The Trustees of Reservations closed the beaches at Wasque, on Chappaquiddick, where waves were breaking over the dunes at high tide. When the storm had passed, David F. Belcher, superintendent of the Trustees' Chappy properties, found some sand fencing knocked down; his crews reset it. Overwash had sent tons of sand onto the dunes, burying the walk-ways; workers pulled them out and repositioned them atop the dunes.

Little more than a day after the storm passed, the beaches were open again. Though they had narrowed during the storm, there was still plenty of room for seagulls, terns, and sandpipers to gorge themselves on the bounty of tiny sea creatures stranded in storm-tossed seaweed. Soon, children were frolicking in the surf as their parents watched from beach chairs, warning them away from the fishermen nearby. It might be a long time before the battered dunes could restore themselves — perhaps a bit inland from where they had been before. But the beach had survived.

People who live on the shore are right to say the weather there seems to have gotten worse. There have been more severe storms since the mid-1980s than there were in the fifties, sixties, and seventies. But wind and water are not what has altered the storm equation on America's coasts. It is the mile after mile of thick new development that has turned natural disturbances into natural disasters. At Wasque, the problem of coastal storms is solved the old-fashioned way — by keeping people and their buildings off the beach. But other coastal areas, such as Sandbridge, rely instead on sea-walls, bulkheads, revetments, groins, and the rest of the arsenal of coastal engineering. In doing so, they are applying technological fixes to problems to which such fixes are not always well-suited. Though there are places where this approach can work, and other places where it must be made to work, overall it is a battle that cannot be won. At some point, the technological fixes employed to prevent shoreline retreat cross the line from resource management to vandalism.

American political institutions, even our national mythology, are ill-suited to the indeterminacy and elasticity of nature. Faced with a problem such as beach erosion, our response is to solve it, not to live with it. It would almost be un-American to concede that it is beyond us, that it is we who must adapt to the ocean, not the other way around. Still less are we willing to accept that some of the things we call natural disasters are part of the process that maintains an ecosystem and helps it function. After years of argument, those who manage the national forests have finally learned this lesson and no longer reflexively rush into action when a lightning strike ignites the underbrush in a national forest. They know a certain amount of burning is natural, and necessary. Even Hurricane Andrew brought some environmental benefits, when it flushed Biscayne Bay, Florida.

Erosion is not a threat to the beach. If sea level is rising, the beach will simply retreat. Where there are no buildings, no human investment, no exploitation of the beach, this retreat occurs almost in silence, and no one calculates it as an economic loss. But when people stake a claim on the coast, citizens and public officials must decide which is more important: preserving the beach or preserving the buildings behind the beach. In the long run, the only practical approach for the first is retreat. The only environmentally acceptable approach to the second, beach renourishment, is incredibly expensive and ultimately futile. People who plan to harden their coast must expect the effort to maintain their armor to become more and more difficult and more and more expensive. In short, they must be prepared to change their plan.

When Europeans first set eyes on the North American coast, many of them viewed it as a forbidding place, dark and mysterious. Soon, though, they felled some trees, set traps for fish, and were making a living from the coast. For the most part, like the Native Americans who preceded them, they lived lightly on the landscape, even through the early years of the twentieth century. Today, most developed coastal areas have trampled the beach under asphalt and concrete, even as they capitalize on the fact that, once, there was an open beach there. Ocean City, Maryland, actually built an official municipal dune, about thirty feet square and surrounded by fencing, in the middle of the city. It is a bizarre reminder of what the place was like before the condos began their march up the coast. Like scores of other places, its gift shops, souvenir stands, and galleries sell merchandise celebrating the region's fishing and seafaring days. But they are exploiting a maritime culture that no longer exists.

In 1995, a subsidiary of Williams Sonoma, the catalogue company for kitchen hobbyists of the upper-middle class, began selling scented bags filled with "beach glass." It was actually broken glass, smoothed in tumbling machines to duplicate the softened edges of the detritus of a functioning beach. But the glass provides the "experience" of beach glass, without the beach traffic. The price: $15 for eight ounces in a bag tied with raffia and an ampule of "refresher oil." The product was perhaps the most blatant acknowledgement yet that the beaches of the United States have become artificial constructs. But there are plenty of others.

Consider a place like Fisher Island, in Miami harbor. The island itself is artificial, built out of dredge spoil. Its beaches are artificially constructed of sand imported from the Bahamas, held in place by groins and other bits of coastal engineering. And it sits near Miami Beach, the most artificial beach of all. Miami Beach long ago abandoned the low-rise vernacular architecture of Florida, with shady porches, shuttered windows, and courtyards for shade. Instead, it has a seemingly endless row of glass towers whose windows do not open and whose air conditioners run constantly, their hum drowning out sound of the surf. From the terraces of the art deco restaurants of South Beach, it is barely possible to see the sand and certainly impossible to hear the waves break—the sound is muffled by the engine noise from hundreds of cars cruising the avenue. Further north, one can drive a six-lane highway past mile upon mile of beachfront condos and never realize there is a beach behind them.

An instructor at the Art Center College of Design in Pasadena, California, has designed an innovative solution to this awkward problem. His plan calls for video equipment, mounted on the ocean-side of a building, to photograph the water and then display the images on a "video wall" that faces the street. In effect, the image of the sea would pass through the building—with the added advantage that the image could be altered to improve upon nature. For example, on gray days the video wall could be reprogrammed to project a view of the beach on a bright afternoon. The only thing missing is the commercial message.

One spring not too long ago, Pilkey took a class of Duke University geology students to view the beaches of Nags Head, which every year loses a few more houses to the surf. At one stop, threatened houses were surrounded with mounds of sandbags while others, their sandbag walls having failed, were leaning crazily onto the open beach, surrounded by damaged decking, ripped boards, and other rubble.

Pilkey pointed to the black geotextile sandbags, some of them already shredding. "Look at these ugly things, on this beautiful beach," he fumed. "This is happening to something that belongs to all of us. Your grandchildren should have this beach, but it should be big and it should be uncluttered."

Nearby, a brick house had collapsed onto the sand. A student surveying its rubble pulled something out of the debris. It was a sign, and he propped it up against the sagging house. It said FOR RENT.

Several hundred miles away and a few months later, Parkinson stood on the bridge over Sebastian Inlet and looked to the south, where houses were creeping their way up the coast. "It is not to late for this stretch of coast," he said. "Right now, we're trying to get the setbacks there. You get a low enough density and you get that setback and you will limit the quickness with which problems arise."

But people are building without permits or getting permits and ignoring their terms. "You can file a permit and then violate it and you know there's no staff on the regulatory agency to catch you," he said sadly. "Eventually this will all be developed," he went on, pointing down the coast, where new rooftops were poking through the trees. "It will be constantly threatened and it will suck resources out of the state."

Advocates of beach nourishment and other armor or stabilization of the coastline call it "shore protection" and say it is the same kind of infrastructure support communities offer, with barely a murmur, to schools, roads, and the like. They say opponents of beach projects are often motivated not by sound reasoning but by unwarranted resentment of people who own property at the beach.

Much of what they say is correct.

But there will be no schools or water lines unless people build and maintain them. That is not the case with the beach. Viewed as infrastructure, the beach survives quite well with hardly any help at all. The beach is not in danger from nature. A natural shoreline does not need "protection." It may change shape or position in response to storms, but that is not a problem unless someone has built something too close to the edge. There is no erosion "problem" until people get into the act. But people are in the act and the problems they cause can only be expected to worsen if sea level continues to rise.

About the time that Hurricane Felix was battering the East Coast, erosion suddenly accelerated on Chappaquiddick, dramatically changing the shape of its southeast corner. No one knew exactly why, but for some reason sand was moving offshore and collecting in a growing shoal. People feared one thin barrier would break through, opening the ocean to a small body of water they called Swan Pond, after the pair of birds that nested there. In the summer of 1997, Chris Kennedy, the superintendent of the Trustees' properties on Cape Cod, Martha's Vineyard, and Nantucket, came to a meeting of islanders to discuss the situation. At first, his words were hardly reassuring. In two years, nine hundred acres had been lost to the ocean, he said.

"When this first started happening, we thought 'Oh, this is awful,'" Kennedy told the group. "But we try to keep things in perspective." He displayed enlargements of aerial photos taken of Wasque in the 1940s. If anything, the island then had been even smaller. There was no Swan Pond. "The beach comes and goes," he said. "It will come back. Maybe not next year or the year after, but sooner or later Wasque Point will return."

NOTES

PREFACE

1. Joseph Slight, a reporter for the *New Bedford Standard Times*, quoted in Everett S. Allen, *A Wind to Shake the World* (Boston: Little, Brown, 1976).

1. SEPTEMBER, REMEMBER

1. Isaac Monroe Cline, *A Century of Progress in the Study of Cyclones, Hurricanes, and Typhoons* (New Orleans: Rogers Printing Co., 1942), p. 26. (President's Address, American Meteorological Society, Pittsburgh meeting, December 29, 1934)
2. David G. McComb, *Galveston: A History*, 2d ed. (Austin: University of Texas Press, 1991), p. 123.
3. Cline, *Century of Progress*, pp. 27–28.
4. Ibid., pp. 29–31.
5. Murat Halstead, *Galveston: The Horrors of a Stricken City* (American Publishers' Association, 1900), p. 78.
6. McComb, *Galveston*, p. 142.
7. Ibid.
8. *Galveston News.*
9. McComb, *Galveston*, p. 142.
10. Estimates of the dead range from eight to twelve.
11. Peter Benchley, *Ocean Planet* (New York: Abrams, 1995), p. 147.
12. *America's Coasts: Progress and Promise* (Coastal States Organization, 1985).
13. Rhode Island Sea Grant, *NOAA's Coastal Ocean Program: Science for Solutions* (Washington, D.C.: National Oceanographic and Atmospheric Administration, Coastal Ocean Office, 1992). This figure excludes Alaska.

2. THE GREAT BEACH

1. In 1998 Leatherman became the director of the International Hurricane Center in Miami, Florida. Meanwhile, his list of the "ten best" beaches in the United States, initially an off-the-cuff evaluation offered to a reporter, has turned into a major sideline. He has begun offering yearly lists, which are widely publicized, and has collected his notes on the nation's best beaches in a book he published himself in 1998. He now has business cards identifying him as "Dr. Beach."

2. Orrin H. Pilkey, et al. *Saving the American Beach: A Position Paper by Concerned Coastal Geologists* (Savannah, Ga.: Skidaway Institute of Oceanography, 1981). The Skidaway Institute of Oceanography is a research unit of the University of Georgia.

3. U.S. Coast and Geodetic Survey, *Report of the Superintendent Showing the Progress of the Work During the Fiscal Year Ending with June 1889* (Washington, D.C.: United States Government Printing Office, 1890), p. 23.

4. Ibid., p. 409.

5. Stephen P. Leatherman, Graham Giese, and Patty O'Donnell, *Historical Cliff Erosion of Outer Cape Cod* (National Park Service, 1981).

6. U.S. Coast and Geodetic Survey, *Report*, p. 404.

7. Leatherman, Giese, and O'Donnell, *Historical Cliff Erosion*, p. 10.

8. Orrin Pilkey and Wallace Kaufman, *The Beaches Are Moving* (Garden City, N.Y.: Anchor Press/Doubleday, 1979), p. 38.

9. The continental shelf of the East Coast is traversed by undersea sand ridges, the remains of ancient barriers. When they were discovered in the 1920s, scientists thought the ridges and swales, parallel to the coastline, were ancient riverbeds. David B. Duane et al., "Linear Shoals on the Atlantic Inner Continental Shelf, Florida to Long Island" in Donald J. P. Swift et al., eds., *Shelf Sediment Transport: Process and Pattern* (Stroudsburg, Pa.: Dowden, Hutchinson, and Ross, 1972), pp. 447–498.

10. Gary Griggs and Lauret Savoy, eds., *Living with the California Coast* (Durham, N.C.: Duke University Press, 1985) p. 34.

11. Admont Clark, *Lighthouses of Cape Cod—Martha's Vineyard—Nantucket* (East Orleans, Mass.: Parnassus Imprints, 1981), p. 33.

12. Ibid., p. 35.

13. Graham Giese, "Cyclical Behavior of the Tidal Inlet at Nauset Beach, Chatham, Massachusetts," in D. G. Aubrey and L. Weishar, eds., *Hydrodynamics and Sediment Dynamics of Tidal Inlets* (New York: Springer-Verlag, 1988), p. 280.

14. Timothy J. Wood, *Breakthrough: The Story of Chatham's North Beach* (Chatham, Mass.: Hyora Publications, 1991), p. 23.

3. ARMOR

1. Joseph M. Heikoff, *Politics of Shore Erosion: Westhampton Beach* (Ann Arbor, Mich.: Ann Arbor Science Publishers, 1976), p. 41.

2. Ibid., p. 43.

3. *New York Times*, December 30, 1994.

4. John Rather, "Coastal Geologist Questions Plan for Restoration of Barrier Beach," *New York Times*, March 24, 1996, sec. 8 (Long Island), p. 8.

5. "Preserving Long Island's Coastline: A Debate on Policy," in *Proceedings of the Fifth Annual Conference of the Long Island Coastal Alliance, Hauppauge, N.Y., April 22, 1994* (Coastal Reports, Inc., 1995).

6. Philip P. Farley, "Coney Island Public Beach and Boardwalk Improvement," *The Municipal Engineers Journal* 9 (1923): 136.9–136.10.

7. Philadelphia District, U.S. Army Corps of Engineers, *P4 Technical Review Submission, New Jersey Shore Protection Study (Townsends Inlet to Cape May Inlet)*, p. 74 ff.

8. Orrin H. Pilkey and E. Robert Thieler, "Artificial Reefs Do Not Work," *The Press* (Atlantic City, N.J.), November 1, 1994, p. A4.

9. Robert G. Dean, *Independent Analysis of Beach Changes in the Vicinity of the Stabeach System at Sailfish Point, Florida: Second Report July 1988 to April 1990*, prepared for Coastal Stabilization, Inc., Rockaway, N.J.

10. *Methodology for Evaluating the Performance of Shore Protection Methods* (Washington, D.C.: Marine Board, National Research Council, 1996).

11. *Beach Nourishment and Protection* (Washington, D.C.: Marine Board, National Research Council, 1995), p. 92.

12. J. Richard Weggel, "Seawalls: The Need for Research, Dimensional Considerations, and a Suggested Classification," *Journal of Coastal Research* special issue 4 (Autumn 1988): 29–39.

13. Quoted in *Duke Research*, 1992, p. 60.

14. *Managing Coastal Erosion* (Washington, D.C.: Marine Board, National Research Council, 1990), p. 59.

15. Mary Jo Hall and Orrin H. Pilkey, "Effects of Hard Stabilization on Dry Beach Width for New Jersey," *Journal of Coastal Research* 7, no. 3 (Summer 1991): 771–785.

16. Weggel, "Seawalls," p. 32.

17. Orrin H. Pilkey and Howard L. Wright 3d, "Seawalls Versus Beaches," *Journal of Coastal Research* special issue 4 (Autumn 1988): 51 ff.

18. Hall and Pilkey, p. 776.

19. William J. Neal et al., *Living with the South Carolina Coast* (Durham, N.C.: Duke University Press, 1984), pp. 114–115.

20. Thomas J. Schoenbaum, *Islands, Capes, and Sounds* (Winston-Salem, N.C.: John F. Blair, 1982), p. 161.

21. P. J. Godfrey, *Oceanic Overwash and Its Ecological Implications on the Outer Banks of North Carolina* (National Park Service Report, 1970).

22. Robert Dolan, Harry Lins, and William E. Odum, "Man's Impact on the Barrier Islands of North Carolina," *American Scientist* 61 (March/April 1973): 152–154.

23. Ibid., pp. 159–160.

24. Robert Dolan and Harry Lins, *The Outer Banks of North Carolina*, U.S. Geological Survey Professional Paper 1177-B (Washington, D.C.: United States Government Printing Office, 1986), p. 38.

25. Weggel, "Seawalls," p. 31.

26. *Rhode Island Development Council Interim Report, Hurricane Rehabilitation Study*, October 1954.

4. UNKIND CUTS

1. Bert and Margie Webber, *Bayocean: The Oregon Town that Fell Into the Sea* (1989; reprint, Medford, Ore., Webb Research Group, 1992), pp. 73–74.

2. Ibid., p. 75.

3. Henry L. Whiting, *Recent Changes in the South Inlet Into Edgartown Harbor, Martha's Vineyard*, Report of the Superintendent, U.S. Coast and Geodetic Survey, Appendix No. 14 (Fiscal Year Ending June, 1889), (Washington, D.C.: United States Government Printing Office, 1890).

4. Robert Dolan and Robert Glassen, "Oregon Inlet, North Carolina: A History of Coastal Change," *Southeastern Geographer* 13, no. 1, p. 9.

5. Karl Nordstrom et al., *Living with the New Jersey Shore* (Durham, N.C.: Duke University Press, 1986), p. 38.

6. Ibid., p. 39.

7. Mindy Fetterman, "Bertha Whips Up Island Debates," *USA Today*, July 15, 1996, p. 3A.

8. Tony Boylan, "Brevard Barks at Erosion's Bite," *Florida Today*, September 24, 1995, p. 1A.

9. Erik J. Olsen, "Beach Management Through Applied Inlet Management," paper presented at 1993 Beach Preservation Technology Conference sponsored by Florida Shore and Beach Preservation Association, pp. 9–10.

10. No relation to the author.

11. Peter Benchley, *Ocean Planet* (New York: Abrams, 1995), p. 146.

5. UNNATURAL APPETITE

1. Susan Miller, "You Want to Stroll the Beach?—Just Try It," *Miami Herald*, May 25, 1970, p. 1.

2. Jay Clarke, "The Sands of Time Are Running Out for Miami Beach," *New York Times*, May 10, 1970, p. 51.

3. Like Professor Dean of the University of Florida, Edwin B. Dean is no relation to the author.

4. Committee on Beach Nourishment and Protection, Marine Board, Commission on Engineering and Technical Systems, National Research Council, "Beach Nourishment and Protection" (Washington, D.C.: National Academy Press, 1995), pp. 168–169.

5. Philip P. Farley, "Coney Island Public Beach and Boardwalk Improvement," *The Municipal Engineers Journal* 9 (1923): 136.3.

6. Ibid., p. 136.18.

7. Ibid., pp. 136.19–20

8. Ibid., p. 136.19.

9. James R. Houston, "Beachfill Performance," *Shore and Beach* (July 1991).

10. Orrin H. Pilkey, "Another View of Beachfill Performance," *Shore and Beach* (April 1992).

11. Committee on Beach Nourishment and Protection, "Beach Nourishment," p. 179.

12. Orrin H. Pilkey et al., "Predicting the Behavior of Beaches: Alternatives to Models," *Littoral 94* (September 26–29, 1994).

13. Shoreline Protection and Beach Erosion Control Task Force, U.S. Army Corps of Engineers, "Shoreline Protection and Beach Erosion Control Study" (Washington, D.C.: Office of Management and Budget, 1994).

14. Anthony DePalma, "Pumping Sand: New Jersey Starts to Replenish Beaches," *New York Times*, February 27, 1990.

15. Jon Nordheimer, "As a Beach Erodes, So Does Faith in Costly Restoration Project," *New York Times*, November 18, 1995, p. B1.

16. ———, "U.S. Says Beach Restoration Will Need a Lot More Sand," *New York Times*, August 29, 1995.

17. Andy Newman, "Effort to Save Strip of Beach May Not Work, Engineer Says," *New York Times*, August 2, 1996, p. B1

18. Jerry Allegood, "Town Considers Assessments to Restore Beach," *Raleigh News and Observer*, October 25, 1996, p. 3A.

19. ———, "Out to Sea," *The Economist*, June 20, 1998.

20. Mike Clary, "Miami Beach's Shoreline Under Siege," *Los Angeles Times* (Washington ed.).

21. Rachel Carson, *The Edge of the Sea* (1955; reprint, Boston: Houghton Mifflin, 1983), p. 140.

22. The author was walking this beach one day, marveling at its hardness, when she spotted a galvanized nail about four inches long lying on the sand. Curious, she picked it up and tried to push it into the sand. It penetrated about a quarter of an inch.

6. CAUSE AND EFFECT

1. Wilson, Scott, "In the Shadow of Cliffs Lies a Shadow of Ruin," *New York Times*, March 10, 1993, p. A14.

2. Gary Griggs and Lauret Savoy, *Living with the California Coast* (Durham, N.C.: Duke University Press, 1985), p. 14.

3. Gerald G. Kuhn and Francis P. Shepard, *Sea Cliffs, Beaches, and Coastal Valleys of San Diego County* (1984; reprint, Berkeley: University of California Press, 1991), p. 52.

4. Griggs and Savoy, *Living*, p. 21.

5. Prentiss Williams, "Los Angeles River Overflowing With Controversy," *California Coast & Ocean* (Summer 1993): 10.

6. Kuhn and Shepard, *Sea Cliffs*, p. 46.

7. Reinhard E. Flick, "The Myth and Reality of Southern California Beaches," *Shore and Beach* (July 1993): 4.

8. Ibid., p. 9.

9. David C. Slade et al., "Putting the Public Trust Doctrine to Work: The Application of the Public Trust Doctrine to the Management of Lands, Waters, and Living Resources of the Coastal States" (Connecticut Department of Environmental Protection, November 1990).

10. *Orion v Washington*, 109 Wash, 2d 621, 747 p.2d 1062 (1987), cited in *Washington Sea Grant Marine Research News*, January 31, 1991.

11. Seth Mydans, "City of Angels Makes Peace in Water Wars," *New York Times*, October 3, 1994, p. A10.

12. *Phillips Petroleum Co. v Mississippi*, 108 S.Ct. 791 (1988).

13. *Marks v Whitney*, (1971) 6 Cal.3d 251.

14. Katherine Stone and Benjamin Kaufman, "Sand Rights: A Legal System to Protect the 'Shores of the Sea,'" *Shore and Beach* 56, no. 3 (July 1988): 7–14.

15. Katherine Stone, "The Use of 'Sand Rights' to Fund Beach Erosion Control Projects," paper presented at 1989 California Shore and Beach Preservation Association Annual Conference, p. 3.

16. Douglas Inman, "Dammed Rivers and Eroded Coasts," in *Critical Problems Relating to the Quality of California's Coastal Zone*, background papers for California Academy of Sciences workshop, January 12–13, 1989.

17. Frank Clifford, "Grand Canyon Flood," *Los Angeles Times*, March 3, 1996.

18. Gregory McNamee, "After the Flood," *Audubon* (September/October 1996).

19. Tom Kenworthy, "River Flow Limits in Grand Canyon Made Permanent," *Washington Post*, October 10, 1996, p. A16.

20. Clifford, "Grand Canyon Flood."

21. ——, "Pumping Life Back Into the Grand Canyon," *Los Angeles Times*, March 27, 1996.

7. THE BIG ONE

1. Robert McNamara, *In Retrospect* (New York: Times Books, 1995), p. 161.

2. 19%, 20%, and 19%, moving north to south.

3. 34%, 30%, and 30%, moving north to south.

4. Edward N. Rappaport, "Preliminary Report Hurricane Andrew 16–28," August 1992, Coral Gables, Fla., National Hurricane Center (updated December 10, 1993).

5. Lawrence S. Tait, ed., "Coastline at Risk: The Hurricane Threat to the Gulf and Atlantic States," Proceedings of the Fourteenth Annual National Hurricane Conference, April 8–10, 1992, Norfolk, Virginia.

6. Robert Sheets, "Stormy Weather," in Lawrence S. Tait, ed., "Hurricanes . . . Different Faces in Different Places" (Atlantic City, N.J.: Proceedings of the Seventeenth Annual National Hurricane Conference, April 11–14, 1995), p. 56.

7. Mark E. Leadon, "Hurricane Opal: Erosional and Structural Impacts to Florida's Gulf Coast," *Shore and Beach* 64, no. 4 (October 1996): 5–14.

8. National Oceanographic and Atmospheric Administration (Coastal Ocean Office), "Coastal Ocean Program: Science for Solutions" (Washington, D.C., 1992), p. 10.

9. Sheets, "Stormy Weather," p. 57.

10. Ibid., p. 61.

11. Robert Dolan and Harry Lins, *The Outer Banks of North Carolina*, U.S. Geological

Survey Professional Paper 1177-B (Washington, D.C.: United States Government Printing Office, 1986), p. 23.

12. S. Jeffess Williams, "Geomorphology and Coastal Processes Along the Atlantic Shoreline, Cape Henlopen, Delaware to Cape Charles, Virginia," paper presented at April 16, 1994, Assateague Shore and Shelf Conference.

13. Robert Dolan, Harry Lins, and Bruce Hayden, "Mid-Atlantic Coastal Storms," *Journal of Coastal Research* 4, no. 3 (Summer 1988): 418.

14. Robert Dolan and Robert E. Davis, "Rating Northeasters," *Marine Weather Log* (Winter 1992): 4–11.

15. Gerald G. Kuhn and Francis P. Shepard, *Sea Cliffs, Beaches, and Coastal Valleys of San Diego County* (1984; reprint, Berkeley: University of California Press, 1991), p. 29, citing R. H. Dana's *Two Years Before the Mast*, Dana's account of his work on boats carrying cattle hides along the coast.

16. Kuhn and Shepard, *Sea Cliffs*, p. 167.

17. Paul J. Herbert, Jerry D. Jarrell, and Max Mayfield, "The Deadliest, Costliest, and Most Intense United States Hurricanes of This Century" (Coral Gables, Fla.: National Oceanographic and Atmospheric Administration Technical Memorandum NWS NHC-31, 1995).

18. "Building Performance: Hurricane Andrew in Florida" (Washington, D.C.: Federal Emergency Management Agency, 1992), p. 37.

19. Ibid.

20. Michael Quint, "A Storm Over Housing Codes," *New York Times*, December 1, 1995, P. D1.

21. "Building Performance," p. 75.

8. CLUES

1. Stephen E. Ambrose, *D-Day, June 6, 1944: The Climactic Battle of World War II* (New York: Simon and Schuster, 1994), p. 74.

2. Ibid., pp. 73–76.

3. Willard Bascom, *Waves and Beaches: The Dynamics of the Ocean Surface* (Garden City, N. Y.: Anchor Press/Doubleday, 1980), p. 231.

4. Ibid., pp. 233–234.

5. Peter A. Howd, and William A. Birkemeier, "Storm-Induced Morphology Changes During Duck85," in Nicholas C. Kraus, ed., *Coastal Sediments '87*. New York: American Society of Civil Engineers, 1987, p. 846.

6. J. R. Allen et al., "A Field Data Assessment of Contemporary Models of Beach Cusp Formation," *Journal of Coastal Research* 12, no. 3 (Summer 1996): 622, 629.

7. Gary Griggs et al., "The Effects of Storm Waves of 1995 on Beaches Adjacent to a Long-Term Seawall Monitoring Site in Northern Monterey Bay, Calif.," *Shore and Beach* 64, no. 1 (January 1996): 34–39.

8. Robert L. Wiegel, "Coastal Engineering Trends and Research Needs," in Billy L.

Edge ed., *Proceedings of Twenty-First Coastal Engineering Conference* (New York: American Society of Civil Engineers, 1988), p. 1.

9. CONSTITUENCY OF IGNORANCE

1. Lorance Lisle and Robert Dolan, "Coastal Erosion and the Cape Hatteras Lighthouse," *Environ Geol Water Sci* 6, no. 3 (1984): 141–146.
2. Harker's Island, N.C., United Methodist Women, *Island Born and Bred* (1987; reprint, Morehead City, N.C.: Herald Printing, 1991), p. 3.
3. Henry Beston, *The Outermost House* (New York: Ballantine, 1971), p. 7.
4. Ibid., p. 230.
5. Coastal Engineering Research Board, *Proceedings of the Sixty-first Meeting of the Coastal Engineering Research Board* (Washington, D.C.: U.S. Army Corps of Engineers, May 10, 1995), p. 22.
6. Bill Emerson and Ted Stevens, "Natural Disasters: A Budget Time Bomb," *Washington Post*, October 31, 1995, p. A13.
7. Joseph B. Treaster, "Insurer Curbing Sales of Policies in Storm Areas," *New York Times*, October 1, 1996, p. A1.
8. Leslie Scism and Martha Brannigan, "Florida Homeowners Find Insurance Pricey, If They Find It at All," *Wall Street Journal*, July 12, 1996, p. A1; Mireya Navarro, "Florida Facing Crisis in Insurance," *New York Times*, April 25, 1996, p. A16.
9. Rutherford Platt, *Land Use and Society* (Washington, D.C.: Island Press, 1996), p. 431.
10. Anna Griffin and Jack Horan, "Teary Residents Shocked at Decimation at Topsail Beach," *Charlotte Observer*, September 8, 1996, p. A16.
11. Robert Dolan, "The Ash Wednesday Storm of 1962: 25 Years Later," *Journal of Coastal Research* 3, no. 2 (1987): ii–iii.
12. Peter Benchley, *Ocean Planet* (New York: Abrams, 1995), p. 147.
13. *America's Coasts: Progress and Promise* (Coastal States Organization, 1985).
14. *Coastal Ocean Program: Science for Solutions* (Washington, D.C.: National Oceanographic and Atmospheric Administration [Coastal Ocean Office], 1992), p. 4.
15. David Owens, *The Coastal Management Plan in North Carolina* (Raleigh: Department of Environmental, Health, and Natural Resources) p. 29.
16. Barry Yeoman, "Shoreline for Sale," *Coastal Review* 14, no. 2 (Summer 1996), p. 8.
17. Craig Whitlock, "Fighting Their Watery Fate," *Raleigh News and Observer*, March 6, 1998, p. A1.
18. Stanley R. Riggs, "Conflict on the Not-So-Fragile Barrier Islands," *Geotimes* 41, no. 12 (December 1996): 15.
19. John A. Humbach, "Private Property vs. Civic Responsibility," paper presented at National Hurricane Conference, Atlantic City, N.J., April 13, 1995.
20. Robert Costanza et al., "Valuation and Management of Wetland Ecosystems," *Ecological Economics* 1 (1989): 335–361, and Robert Costanza and Charles Perrings, "A Flexible Assurance Bonding System for Improved Environmental Management," *Ecological Economics* 2 (1990): 57–75.

21. James R. Houston, "International Tourism and U.S. Beaches," *The CERCular* 96, no. 2 (June 1996).
22. Mark Crowell, presentation at the Second Thematic Conference on Remote Sensing for Marine and Coastal Environments, 1993.

10. FOR SALE

1. *Vineyard Gazette*, July 10, 1987.
2. Jason Gay, "Private Beach Access Costs Are Climbing," *Vineyard Gazette*, July 5, 1996, p. 1D.
3. Gay, "Dunkirk at South Beach," *Vineyard Gazette*, July 14, 1987.
4. Dorsey Griffith, "Furor Rises on Sale of South Beach, Developers Warned of Legal Action," *Vineyard Gazette*, July 24, 1987, p. 1.
5. Griffith, "Edgartown Regains South Beach Control in Court Suit to Block Private Developers," *Vineyard Gazette*, July 31, 1987, p. 1.
6. Griffith, "Town Rejects Offer from Beach Buyers," *Vineyard Gazette*, October 2, 1987, p. 1.
7. Griffith, "The Story of South Beach: Victory for Town and State," *Vineyard Gazette*, May 12, 1988.
8. Griffith, "James Gutensohn Speaks Out Over South Beach," *Vineyard Gazette*, September 8, 1987.
9. David Stick, *The Outer Banks of North Carolina* (Chapel Hill: University of North Carolina Press, 1958), p. 250.
10. Ruthe Wolverton and Walt Wolverton, *The National Seashores* (Kensington, Md.: Woodbine House, 1988), p. 43.
11. John McPhee, *Encounters with the Arch-Druid* (New York: Noonday, 1971), pp. 77–150.
12. Wolverton and Wolverton, *National Seashores*, pp. 166–170.
13. Joseph J. Thorndike, *The Coast: A Journey Down the Atlantic Shore* (New York: St. Martin's, 1993), pp. 134–135.
14. Wolverton and Wolverton, *National Seashores*, p. 69.
15. Ibid., p. 192 ff.
16. Peter Lord, "Rare Earth: Block Island Wins Recognition as One of Hemisphere's 'Last Great Places,' " *Providence Sunday Journal*, May 12, 1991, p. A1.
17. Ibid.
18. Carla Hall, "Fund-Raising Frenzy Buys Santa Barbara Park," *Los Angeles Times* (Washington ed.), March 12, 1996, p. B2.
19. *Vineyard Gazette*, July 5, 1996.
20. Thomas Farragher, "Summer Traffic Crunch Lifts Cape Land-Bank Campaign," *Boston Sunday Globe*, July 19, 1998, p. B1.
21. "Tidal Pool of Tears," *Vineyard Gazette*, July 24, 1987.
22. Marilyn W. Thompson, "Wild Horses Champ at Tourist Bit," *Washington Post*, August 30, 1996, p. A1.

23. Duane De Freese, "Protecting Coastal Diversity: A Local Perspective," *Coastal Currents* 2, no. 3 (Summer 1994): 12–13.

24. Margaret Greer, *The Sands of Time: A History of Hilton Head Island* (Hilton Head, S.C.: South/Art, 1989).

25. Gunnar Hansen, *Islands at the Edge of Time* (Washington, D.C.: Island Press, 1993).

26. *Coastal Heritage* 6, no. 1 (Winter 1991): 5–6.

BIBLIOGRAPHY

Except as noted, statements quoted in this book were made in interviews with or at presentations attended by the author.

PREFACE

Allen, Everett S. *A Wind to Shake the World*. Boston: Little, Brown, 1976.

"Complete Historical Record of New England's Stricken Area." *New Bedford Standard Times*, September 21, 1938.

"The Hurricane of '38." *New Bedford Standard Times*, September 21, 1988.

1. SEPTEMBER, REMEMBER

Baker, T. Lindsay. "Galveston Graderaising Was a Big Engineering Project." *Galveston Daily News*, May 11, 1974, p. 8A.

Cline, Isaac Monroe. "A Century of Progress in the Study of Cyclones, Hurricanes, and Typhoons." New Orleans: Rogers Printing Co., 1942. (President's Address, American Meteorological Society, Pittsburgh Meeting, American Association for the Advancement of Science, December 29, 1934).

Coastal Ocean Program: Science for Solutions. Washington, D.C.: National Oceanographic and Atmospheric Administration (Coastal Ocean Office), 1992.

Department of the Army, Waterways Experiment Station, Corps of Engineers, Coastal Engineering Research Center. *Shore Protection Manual* (vols. 1 and 2). Washington, D.C.: 1984.

Eisenhour, Virginia, *Galveston—A Different Place*. 5th ed. Galveston, Texas, 1991.

Green, Nathan C., ed. *Story of the Galveston Flood*. Baltimore: Woodward, 1900.

Halstead, Murat. *Galveston: The Horrors of a Stricken City*. American Publishers' Association, 1900.

Hansen, Gunnar. *Islands at the Edge of Time*. Washington, D.C.: Island Books, 1993.

McComb, David G. *Galveston, A History*. 2d. ed. Austin: University of Texas Press, 1991.

Miller, Ray. *Galveston*. 5th ed. Houston: Gulf Publishing, 1993.

National Research Council. *Managing Coastal Erosion*. Washington, D.C.: National Academy Press, 1990.

Thorndike, Joseph J. *The Coast: A Journey Down the Atlantic Shore*. New York: St. Martin's, 1993.

Weems, John Edward. *A Weekend in September.* 5th ed. College Station: Texas A&M University Press, 1993.

2. THE GREAT BEACH

Beston, Henry. *The Outermost House.* New York: Ballantine, 1979.

Brigham, Albert Perry. *Cape Cod and the Old Colony.* New York: Grosset and Dunlap.

Chamberlain, Barbara Blau. *These Fragile Outposts.* Yarmouthport, Mass.: Parnassus Imprints, 1981.

Clark, Admont. *Lighthouses of Cape Cod—Martha's Vineyard—Nantucket.* East Orleans, Mass. Parnassus Imprints. 1992.

Dolan, Robert, and Harry Lins. *The Outer Banks of North Carolina.* U.S. Geological Survey Professional Paper 1177-B. Washington, D.C.: United States Government Printing Office, 1986.

Duane, David B., et al. "Linear Shoals on the Atlantic Inner Continental Shelf, Florida to Long Island." In Donald J. P. Swift et al., eds., *Shelf Sediment Transport: Process and Pattern.* Stroudsburg, Pa.: Dowden, Hutchinson, and Ross, 1972, pp. 447–498.

Finch, Robert. *Outlands: Journeys to the Outer Edges of Cape Cod.* Boston: David R. Godine, 1986.

Fox, William T. *At the Sea's Edge.* New York: Prentice-Hall, 1983.

Giese, Graham. "Cyclical Behavior of the Tidal Inlet at Nauset Beach, Chatham, Massachusetts." In D. G. Aubrey and L. Weishar, eds., *Hydrodynamics and Sediment Dynamics of Tidal Inlets.* New York: Springer-Verlag, 1988.

Goudie, Andrew. *The Human Impact on the Natural Environment.* Cambridge, Mass.: MIT Press, 1994.

Griggs, Gary and Lauret Savoy, eds. *Living with the California Coast.* Durham, N.C.: Duke University Press, 1985.

Hay, John. *The Great Beach.* New York: Norton, 1963 [1980 ed.].

Klauber, Avery. "Our Disappearing Coastline: Losing More Than Just the Sand on the Beach." *Nor'easter* (Fall 1989).

Komar, Paul. "Ocean Processes and Hazards Along the Oregon Coast." *Oregon Geology* 54, no. 1 (January 1992).

Leatherman, Stephen P. "Time Frames for Barrier Island Migration." *Shore and Beach* 55, nos. 3/4 (July/October 1987): 82–86.

———. *Cape Cod Field Trips.* College Park, Md.: Laboratory for Coastal Research, University of Maryland, 1988.

Leatherman, Stephen P., Graham Giese, and Patty O'Donnell. "Historical Cliff Erosion of Outer Cape Cod." National Park Service, 1981.

Marindin, H. L. "Encroachment of the Sea upon the Coast of Cape Cod." In *Annual Report, U.S. Coast and Geodetic Survey, 1889.* Appendix 12, pp. 403–407; appendix 13, pp. 409–457.

Nickerson, Joshua Atkins. *Days to Remember.* Chatham, Mass.: The Chatham Historical Society, 1988.

Norris, Robert M. "Sea Cliff Erosion: A Major Dilemma." *California Geology* (August 1990).

Pilkey, Orrin H., et al. "Saving the American Beach: A Position Paper by Concerned Coastal Geologists." Savannah, Ga., Skidaway Institute of Oceanography, 1981.

Pilkey, Orrin, and Wallace Kaufman. *The Beaches Are Moving.* Garden City, N.Y.: Anchor Press/Doubleday, 1979.

Sargeant, William. *Storm Surge.* Hyannis, Mass.: Parnassus Imprints, 1995.

State of the Cape, 1994. Orleans, Mass.: Association for the Preservation of Cape Cod, 1993.

Strahler, Arthur N. *A Geologist's View of Cape Cod.* Orleans, Mass.: Parnassus Imprints, 1966 [1988 ed.].

Thoreau, Henry David. *Cape Cod.* New York: Little, Brown, 1865. Reprint, New York: New York Graphic Society, 1985.

Thorndike, Joseph J. *The Coast: A Journey Down the Atlantic Shore.* New York: St. Martin's, 1993.

U.S. Coast and Geodetic Survey. *Report of the Superintendent Showing the Progress of the Work during the Fiscal Year Ending with June, 1889.* Washington, D.C.: United States Government Printing Office, 1890.

Wood, Timothy J. *Breakthrough: The Story of Chatham's North Beach.* Chatham, Mass.: Hyora, 1991.

Zeigler, John M. *Beach Studies in the Cape Cod Area (August 1953–April 1960).* Woods Hole, Mass.: Woods Hole Oceanographic Institution, 1960.

3. ARMOR

Alexander, John, and James Lazell. *Ribbon of Sand.* Chapel Hill, N.C.: Algonquin, 1992.

"An Assessment of Shoreline Protection Measures along the Central California Coast," University of California, Santa Cruz, E/G-2, 1983–84.

Basco, David R. "Boundary Conditions and Long-Term Shoreline Change Rates for the Southern Virginia Ocean Coastline." *Shore and Beach* 59, no. 4 (October 1991): 8–13.

Basco, David R., et al. "Preliminary Statistical Analysis of Beach Profiles for Walled and Non-Walled Sections at Sandbridge, Virginia." Paper presented at 7th National Conference on Beach Preservation Technology, Tampa, Fla., February 9–11, 1994.

Birkemeier, William, Robert Dolan, and Nina Fisher. "The Evolution of a Barrier Island: 1930–1980." *Shore and Beach* (April 1984): 3–12.

"Cape May County and Monmouth County Field Trip Guides." Assateague Shore and Shelf Workshop Twenty-One (held at Richard Stockton College of New Jersey, Pomona, N.J., April 21–22, 1995).

Committee on Beach Nourishment and Protection (Marine Board, Commission on Engineering and Technical Systems, National Research Council). *Beach Nourishment and Protection.* Washington, D.C.: National Academy Press, 1995.

Department of the Army, Waterways Experiment Station, Corps of Engineers, Coastal Engineering Research Center. *Shore Protection Manual.* Vols. 1 and 2. 4th ed. Washington D.C., 1984.

Dolan, Robert. "Barrier Dune System along the Outer Banks of North Carolina: A Reappraisal." *Science* (April 21, 1972): 286–288.

Dolan, Robert, and Bruce Hayden. "Adjusting to Nature in Our National Seashores." *Environmental Journal* (June 1974): 9–14.

Dolan, Robert, and Harry Lins. *The Outer Banks of North Carolina.* U.S. Geological Survey Professional Paper 1177-B. Washington, D.C.: United States Government Printing Office, 1986.

Dolan, Robert, Harry Lins, and William E. Odum. "Man's Impact on the Barrier Islands of North Carolina." *American Scientist* 61 (March/April 1973): 152–162.

Farley, Philip P. "Coney Island Public Beach and Boardwalk Improvement." *Municipal Engineers Journal* 9 (1923).

Farrell, Stewart. "Preliminary Analysis of the Impact of New Beach Technology on Avalon, New Jersey." Paper presented at Twentieth Assateague Shelf and Shore Workshop, Ocean City, Maryland, April 15–16, 1994.

First Coastal Corporation. "Dune Road Westhampton — Potential for Overwash & Breaching." Westhampton Beach, N.Y., 1992.

Fowler, Jimmy E. *Scour Problems and Methods for Prediction of Maximum Scour at Vertical Seawalls.* Washington, D.C.: Department of the Army, U.S. Army Corps of Engineers, 1992.

Fulton-Bennett, Kim, and Gary B. Griggs. *Coastal Protection Structures and Their Effectiveness.* Marine Sciences Institute, University of California, Santa Cruz and State of California Dept. of Boating and Waterways.

Griggs, Gary B., James F. Tait, and Wendy Corona. "The Interaction of Seawalls and Beaches: Seven Years of Monitoring in Monterey Bay, California." *Shore and Beach* (July 1994).

Hall, Mary Jo, and Orrin H. Pilkey. "Effects of Hard Stabilization Dry Beach Width for New Jersey." *Journal of Coastal Research* 7, no. 3 (Summer 1991): 771–785.

Heikoff, Joseph M. *Politics of Shore Erosion: Westhampton Beach.* Ann Arbor, Mich.: Ann Arbor Science Publishers, 1976.

Leatherman, Stephen P. *Shoreline Changes at Wainscott, East Hampton, New York.* College Park: University of Maryland, Laboratory for Coastal Research, 1989.

Managing Coastal Erosion. Washington, D.C.: Marine Board. National Research Council, 1990.

Marine Board, National Research Council. "Methodology for Evaluating the Performance of Shore Protection Methods." [Statement of research agenda]. Washington, D.C., 1996.

McCormick, Larry R., et al. *Living with Long Island's South Shore.* Durham: Duke University Press, 1984.

Neal, William J., et al. *Living with the South Carolina Shore.* Durham: Duke University Press, 1984.

Philadelphia District, U.S. Army Corps of Engineers. *P4 Technical Review Submission, New Jersey Shore Protection Study (Townsends Inlet to Cape May Inlet).*

Pilkey, Orrin H., and E. Robert Thieler. "Artificial Reefs Do Not Work." *The Press* (Atlantic City, N.J.), November 1, 1994, p.A4.

Pilkey, Orrin H., and Howard L. Wright. "Seawalls Versus Beaches." *Journal of Coastal Research* special issue 4 (Autumn 1988).

"Preserving Long Island's Coastline: A Debate on Policy." *Proceedings, Fifth Annual Conference of the Long Island Coastal Alliance,* Hauppauge, N.Y., April 22, 1994. New York: Coastal Reports Inc., 1995.

Rather, John. "Coastal Geologist Questions Plan for Restoration of Barrier Beach." *New York Times,* March 24, 1996, sec. 8 (Long Island), p. 8.

Schoenbaum, Thomas J. *Islands, Capes, and Sounds.* Winston-Salem, N.C.: John F. Blair, 1982.

Shipman, Hugh, and Douglas J. Canning. *Cumulative Environmental Impacts of Shoreline Stabilization on Puget Sound.* Proceedings, Coastal Zone '93. New York: American Society of Civil Engineers, 1993, pp. 2233–2242.

"State Officials Defend Beachsaver After Charges." *Ocean City Sentinel-Ledger.* November 1994.

Stick, David. *The Outer Banks of North Carolina.* Chapel Hill: University of North Carolina Press, 1958.

Terchunian, A. V., and C. L. Merkert. "Little Pikes Inlet, Westhampton, New York." *Journal of Coastal Research* 11, no. 2 (Spring 1995): 697–703.

Terich, Thomas A. *Structural Erosion Protection of Beaches Along Puget Sound: An Assessment of Alternatives and Research Needs.* Washington Division of Geology and Earth Resources Bulletin 78, 1989.

Thorndike, Joseph J. *The Coast: A Journey Down the Atlantic Shore.* New York: St. Martin's, 1993.

Weggel, J. Richard. "Seawalls: The Need for Research, Dimensional Considerations, and a Suggested Classification." *Journal of Coastal Research* special issue 4 (Autumn 1988): 29–39.

4. UNKIND CUTS

Aubrey, D. G., and P. E. Speer. "Updrift Migration of Tidal Inlets." *Journal of Geology* 92 (1984): 531–545.

Boylan, Tony. "Brevard Barks at Erosion's Bite." *Florida Today,* September 24, 1995, p. 1A.

Clausner, James, et al. "Sand Bypassing at Indian River Inlet, Delaware." U.S. Army Corps of Engineers Waterways Experiment Station. *The CERCular* 92, no. 1 (March 1992): 1–6.

Clausner, James E., David R. Patterson, and Gus Rambo. *Fixed Sand Bypassing Plants: An Update.* Conference Proceedings of Beach Preservation Technology '90. St. Petersburg, Fla.: Shore and Beach Preservation Association, 1990.

Dolan, Robert, and Robert Glassen. "Oregon Inlet, North Carolina: A History of Coastal Change." *Southeastern Geographer* 13, no. 1.

Dorwart, Jeffrey M. *Cape May County, New Jersey*. New Brunswick: Rutgers University Press, 1993.

Fetterman, Mindy. "Bertha Whips Up Island Debates." *USA Today*, July 15, 1996, p. 3A.

Giese, Graham. "Cyclical Behavior of the Tidal Inlet at Nauset Beach, Chatham, Massachusetts." In D. G. Aubrey and L. Weishar, eds., *Hydrodynamics and Sediment Dynamics of Tidal Inlets*. New York: Springer-Verlag, 1988.

Herbst, Joyce. *Oregon Coast*. Portland: Frank Amato, 1985.

Hurley, George M., and B. Suzanne. *Ocean City*. Virginia Beach, Va.: Donning, 1979.

Komar, Paul D. *The Pacific Northwest Coast*. Durham: Duke University Press, in press.

Leatherman, Stephen P., and Jakob J. Møller. "Morris Island Lighthouse: A 'Survivor' of Hurricane Hugo." *Shore and Beach* 59, no. 1 (January 1991): 11–15.

Leatherman, Stephen P., et al. "Shoreline and Sediment Budget Analysis of North Assateague Island, Maryland." In Nicholas C. Kraus, ed., *Coastal Sediments '87*, New York: American Society of Civil Engineers, 1987.

Leatherman, Stephen P., et al. *Vanishing Lands*. College Park, Md.: University of Maryland, 1995.

McBride, Randolph A. "Tidal Inlet History, Morphology, and Stability, Eastern Coast of Florida, USA." In Nicholas C. Kraus, ed., *Coastal Sediments '87*, New York: American Society of Civil Engineers, 1987.

Methot, June. *Up and Down the Beach*. Navesink, N.J.: Whip, 1988.

Nordstrom, Karl F., et al. *Living With the New Jersey Shore*. Durham: Duke University Press, 1986.

Olsen, Erik J. "Beach Management Through Applied Inlet Management." Paper presented at 1993 Beach Preservation Technology Conference sponsored by Florida Shore and Beach Preservation Association.

Pilkey, Orrin H. "Barrier Islands: Formed of Fury, They Roam and Fade." *SeaFrontiers* (December 1990): 30–36.

Quinn, John R. *One Square Mile on the Atlantic Coast: An Artist's Journal of the New Jersey Shore*. New York: Walker, 1993.

Rambo, Gus, and James E. Clausner. "Jet Sand Bypassing, Indian River Inlet, Delaware." *Dredging Research* 89 (November 1989).

Roberts, Russell and Rich Youmans. *Down the Jersey Shore*. New Brunswick: Rutgers University Press, 1993.

Scott, Jane. *Between Ocean and Bay*. Centreville, Md.: Tidewater, 1991.

Terich, Thomas A., and Paul D. Komar. "Bayocean Spit, Oregon: History of Development and Erosional Destruction." *Shore and Beach* 42, no. 2: 3–10.

United States Park Service. *Assateague Island* (Handbook 106). Washington, D.C.: United States Government Printing Office, 1980.

Webber, Bert, and Margie Webber. *Bayocean, The Oregon Town that Fell Into the Sea*. Medford, Ore.: Webb Research Group, 1989.

Whiting, Henry L. *Recent Changes in the South Inlet Into Edgartown Harbor, Martha's Vineyard*. Report of the Superintendent, U.S. Coast and Geodetic Survey, Appendix

No. 14 (Fiscal Year Ending June, 1889). Washington, D.C.: United States Government Printing Office, 1890.

5. UNNATURAL APPETITE

Allegood, Jerry. "Town Considers Assessments to Restore Beach." *Raleigh News and Observer*, October 25, 1996, p. 3A.

Armbruster, Ann. *The Life and Times of Miami Beach*. New York: Knopf, 1995.

Ashley, Gail M., et al. "A Study of Beachfill Longevity, Long Beach Island, N.J." In Nicholas C. Kraus, ed., *Coastal Sediments '87*, New York: American Society of Civil Engineers, 1987.

Clarke, Jay. "The Sands of Time Are Running Out for Miami Beach." *New York Times*, May 10, 1970, p. 51.

Clary, Mike. "Miami Beach's Shoreline Under Siege." *Los Angeles Times* (Washington ed.).

Committee on Beach Nourishment and Protection, Marine Board, Commission on Engineering and Technical Systems, National Research Council. *Beach Nourishment and Protection*. Washington, D.C.: National Academy Press, 1995.

Committee on Sea Turtle Conservation, Board on Environmental Studies and Toxicology, Board on Biology, Commission on Life Sciences, National Research Council. *Decline of the Sea Turtles*. Washington, D.C.: National Academy Press, 1990.

Dean, R. G. "Principles of Beach Nourishment." In Paul D. Komar, ed., *CRC Handbook of Coastal Processes and Erosion*. Boca Raton, Fla.: CRC Press, pp. 217–231.

DePalma, Anthony. "Pumping Sand: New Jersey Starts to Replenish Beaches." *New York Times*, February 27, 1990.

Emmet, Brian K., et al. "Coney Island Storm Damage Reduction Plan." *Shore and Beach* 63, no. 4 (October 1995): 15–24.

Farley, Philip P. "Coney Island Public Beach and Boardwalk Improvement." *The Municipal Engineers Journal* 9 (1923).

Houston, James R. "Beachfill Performance." *Shore and Beach* 59, no. 3 (July 1991): 15–24.

——. "The Economic Value of Beaches." *Coastal Engineering Research Center CERCular* 95, no. 4 (December 1995).

Jones-Jackson, Patricia. *When Roots Die: Endangered Traditions on the Sea Islands*. Athens and London: University of Georgia Press, 1987.

Leatherman, Stephen P. Statement to Maryland State Legislature Committee on Appropriations, Annapolis, Md., January 28, 1992.

May, James P., and Frank W. Stapor Jr. "Beach Erosion and Sand Transport at Hunting Island, South Carolina, U.S.A." *Journal of Coastal Research* 12, no. 3 (Summer 1996): 714–725.

McNinch, Jesse, and John T. Wells. "Effectiveness of Beach Scraping as a Method of Erosion Control." *Shore and Beach* 60, no. 1 (January 1992): 13–20.

Miller, Susan. "You Want to Stroll the Beach?—Just Try It." *Miami Herald*, May 25, 1970, p. 1.

Newman, Andy. "Effort to Save Strip of Beach May Not Work, Engineer Says." *New York Times*, August 2, 1996, p. B1.

Nordheimer, Jon. "Beach Project Pumps Sand and Money." *New York Times*, March 12, 1994.

———. "U.S. Says Beach Restoration Will Need a Lot More Sand." *New York Times*, August 29, 1995.

———."As a Beach Erodes, So Does Faith in Costly Restoration Project." *New York Times*, November 18, 1995, p. B1.

Paquette, Carole. "Fire I. Owners Propose Tax District." *New York Times*, January 21, 1996, p. 13LI.

Pilkey, Orrin H. "Another View of Beachfill Performance." *Shore and Beach* 60, no. 2 (April 1992): 20–25.

Pilkey, Orrin H. Jr., et al. "Living With the East Florida Shore." Durham: Duke University Press, 1984.

Pilkey, Orrin H., et al. "Predicting the Behavior of Beaches: Alternatives to Models." *Littoral 94* (September 26–29, 1994).

Shoreline Protection and Beach Erosion Control Task Force, U.S. Army Corps of Engineers. *Shoreline Protection and Beach Erosion Control Study*. Washington, D.C.: Office of Management and Budget, 1994.

Tait, Lawrence S., ed. *Sand Wars, Sand Shortages & Sand-Holding Structures*. Proceedings, Eighth National Conference on Beach Preservation Technology. Tallahassee, Fla.: Florida Shore and Beach Preservation Association, 1995.

Wells, John T., and McNinch, Jesse. "Beach Scraping in North Carolina with Special Reference to Its Effectiveness During Hurricane Hugo." *Journal of Coastal Research* special issue 8 (1991): 249–261.

Wiegel, Robert L. "Keynote Address: Some Notes on Beach Nourishment." Gainesville, Fla.: Proceedings, Beach Preservation Technology '88. March 23–25, 1988.

6. CAUSE AND EFFECT

Adler, Tina. "Healing Waters." *Science News* 150 (September 21, 1996): 188–189.

Carson, Rachel. *The Edge of the Sea*. 1955. Reprint, Boston: Houghton Mifflin, 1983.

Clifford, Frank. "Grand Canyon Flood." *Los Angeles Times*, March 25, 1996.

———. "Pumping Life Back Into the Grand Canyon." *Los Angeles Times*, March 27, 1996.

Coastal Engineering Research Board. *Proceedings of the Sixty-First Meeting of the Coastal Engineering Research Board*. Washington, D.C.: U.S. Army Corps of Engineers, May 10, 1995.

Flick, Reinhard E. "The Myth and Reality of Southern California Beaches." *Shore and Beach* 61, no. 3 (July 1993): 3–13.

Griggs, Gary B. "California's Retreating Shoreline: The State of the Problem." Proceedings, Coastal Zone '87, New York: American Society of Civil Engineers, 1993, pp. 1370–1383.

Griggs, Gary, and Lauret Savoy. *Living with the California Coast*. Durham: Duke University Press, 1985.

Inman, Douglas L. "Dammed Rivers and Eroded Coasts." In *Critical Problems Relating to the Quality of California's Coastal Zone*. Background papers for California Academy of Sciences workshop, January 12–13, 1989.

Kenworthy, Tom. "River Flow Limits in Grand Canyon Made Permanent." *Washington Post*, October 10, 1996, p. A16.

Kuhn, Gerald G., and Francis P. Shepard. "Beach Processes and Seacliff Erosion in San Diego County, California." In Paul D. Komar, ed., *CRC Handbook of Coastal Processes and Erosion*, Boca Raton, Fla.: CRC Press, pp. 267–284.

———. *Sea Cliffs, Beaches, and Coastal Valleys of San Diego County*. 1984. Reprint, Berkeley: University of California Press, 1991.

Leidersdorf, Craig B., Ricky C. Hollar, and Gregory Woodell. "Beach Enhancement through Nourishment and Compartmentalization: The Recent History of Santa Monica Bay." In *Beach Nourishment Engineering and Management Considerations*. Proceedings, Eighth Symposium on Coastal and Ocean Management, American Shore and Beach Preservation Association/American Society of Civil Engineers, New Orleans, La., July 19–23, 1993, pp. 71–85.

McNamee, Gregory. "After the Flood." *Audubon* (September/October 1996).

Mydans, Seth. "City of Angels Makes Peace in Water Wars." *New York Times*, October 3, 1994, p. A10.

Reisner, Marc, "Cadillac Desert: The American West and Its Disappearing Water." 1986. Reprint, New York: Penguin, 1993.

Slade, David C., et al. "Putting the Public Trust Doctrine to Work: The Application of the Public Trust Doctrine to the Management of Lands, Waters and Living Resources of the Coastal States." Connecticut Department of Environmental Protection, November 1990.

Stone, Katherine, "The Use of 'Sand Rights' to Fund Beach Erosion Control Projects." Paper presented at California Shore and Beach Preservation Association annual conference, 1989.

Stone, Katherine and Benjamin Kaufman. "Sand Rights: A Legal System to Protect the Shores of the Sea." *Shore and Beach* 56, no. 3 (July 1988): 7–14.

Wicinas, David. *Sagebrush and Cappuccino: Confessions of an LA Naturalist*. San Francisco: Sierra Club Books, 1995.

Wiegel, Robert L. "Ocean Beach Nourishment on the USA Pacific Coast." *Shore and Beach* 62, no. 1 (January 1994): 11–36.

Williams, Prentiss. "Los Angeles River Overflowing With Controversy." *California Coast & Ocean* (Summer 1993): 6–18.

Wilson, Scott. "In the Shadow of Cliffs Lies a Shadow of Ruin." *New York Times*, March 10, 1993, p. A14.

Woodell, Gregory, and Ricky Hollar. "Historical Changes in the Beaches of Los Angeles

County." Proceedings, Coastal Zone '91. New York: American Society of Civil Engineers, 1991, pp. 1342–1355.

7. THE BIG ONE

Ballance, Alton. *Ocracokers*. Chapel Hill: University of North Carolina Press: 1989.

Buckley, J. Taylor. "Aid and Insurance Help, but the Victims Pay a Lot." *USA Today*, September 6, 1996, p. 1A.

Dolan, Robert. "The Ash Wednesday Storm of 1962: 25 Years Later." *Journal of Coastal Research* 3, no. 2: ii–vi.

Dolan, Robert, and Robert E. Davis. "Rating Northeasters." *Marine Weather Log* (Winter 1992): 4–11.

Dolan, Robert, and Bruce Hayden. "Storms and Shoreline Configuration." *Journal of Sedimentary Petrology* 51, no. 3 (September 1981): 737–744.

Dolan, Robert, and Harry Lins. *The Outer Banks of North Carolina*. U.S. Geological Survey Professional Paper 1177-B. Washington, D.C.: United States Government Printing Office, 1986.

Dolan, Robert, Harry Lins, and Bruce Hayden. "Mid-Atlantic Coastal Storms." *Journal of Coastal Research* 4, no. 3 (Summer 1988): 417–433.

Federal Emergency Management Agency. "Building Performance: Hurricane Andrew in Florida." Washington, D.C., 1992.

Federal Emergency Management Agency. "Coastal Construction Manual." Washington, D.C., 1985.

Federal Emergency Management Agency. "Manufactured Home Installation in Flood Hazard Zones." Washington, D.C., 1985.

Gray, William M., et al. "Long Period Variations in African Rainfall and Hurricane Related Destruction Along the US East Coast." Paper presented at Fourteenth Annual National Hurricane Conference, Norfolk, Va., April 10, 1992.

Hayden, Bruce. "Secular Variation in Atlantic Coast Extratropical Cyclones." *Monthly Weather Review* 109, no. 1 (January 1981).

Herbert, Paul J., Jerry D. Jarrell, and Max Mayfield. "The Deadliest, Costliest, and Most Intense United States Hurricanes of This Century." Coral Gables, Fla.: National Oceanographic and Atmospheric Administration, Technical Memorandum NWS NHC-31, 1995.

Holleran, Michael, Michael Everett, and Judith Benedict. *Building at the Shore*. Providence, R.I., Rhode Island Department of Coastal Management (undated).

Kuhn, Gerald G., and Francis P. Shepard. "Should Southern California Build Defenses Against Violent Storms Resulting in Lowland Flooding As Discovered in Records of Past Century." *Shore and Beach* (October 1981).

——. *Sea Cliffs, Beaches, and Coastal Valleys of San Diego County*. 1984. Reprint, Berkeley: University of California Press, 1991.

Leadon, Mark E. "Hurricane Opal: Erosional and Structural Impacts to Florida's Gulf Coast." *Shore and Beach* 64, no. 4 (October 1996): 5–14.

Mayer, Caroline E. "Withstanding a Huff and a Puff." *Washington Post*, August 31, 1996. p. E1.

National Oceanographic and Atmospheric Administration (Coastal Ocean Office). *Coastal Ocean Program: Science for Solutions*. Washington, D.C., 1992.

Neuman, Charles J. *Tropical Cyclones of the North Atlantic Ocean, 1871–1986* (with additions). 3d rev. Asheville, N.C.: National Climatic Data Center in Cooperation with National Hurricane Center, 1987.

Nichols, Robert J., et al. "Erosion in Coastal Settings and Pile Foundations." *Shore and Beach* 63, no. 4 (October 1995): 11–17.

Platt, Rutherford, et al., eds. *Cities on the Beach*. Chicago: University of Chicago Press, 1987.

Quint, Michael. "A Storm Over Housing Codes." *New York Times*, December 1, 1995, p. D1.

Rappaport, Edward N. *Preliminary Report Hurricane Andrew, 16–28 August 1992*. Coral Gables, Fla.: National Hurricane Center (updated December 10, 1993).

Tait, Lawrence S., ed. "Coastline at Risk: The Hurricane Threat to the Gulf and Atlantic States." Excerpts from the Fourteenth Annual National Hurricane Conference, Norfolk, Virginia, April 8–10, 1992.

——. "Hurricanes . . . Different Faces in Different Places." Proceedings, Seventeenth Annual National Hurricane Conference, April 11–14, 1995, Atlantic City, N.J.

Williams, S. Jeffess. "Geomorphology and Coastal Processes Along the Atlantic Shoreline, Cape Henlopen, Delaware to Cape Charles, Virginia." Paper presented at the Assateague Shore and Shelf Conference, April 16, 1994.

8. CLUES

Allen, J. R., et al. "A Field Data Assessment of Contemporary Models of Beach Cusp Formation." *Journal of Coastal Research* 12, no. 3 (Summer 1996): 622–629.

Ambrose, Stephen E. *D-Day, June 6, 1944: The Climactic Battle of World War II*. New York: Simon and Schuster, 1994.

Baird, Andrew J., and Diane P. Horn. "Monitoring and Modeling Groundwater Behavior in Sandy Beaches." *Journal of Coastal Research* 12, no. 3: 630–640.

Basco, David R. "Preliminary Statistical Analysis of Beach Profiles for Walled and Non-walled Sections at Sandbridge, Virginia." Paper presented at Twentieth Assateague Shelf and Shore Workshop, Ocean City, Maryland, April 15–16, 1994.

Bascom, Willard. *Waves and Beaches: The Dynamics of the Ocean Surface*. New York: Anchor Press/Doubleday, 1980.

Birkemeier, William A. *The Duck94 Nearshore Experiment*. Vicksburg, Miss.: U.S. Army Corps of Engineers, 1994.

Bokuniewicz, H., and J. Tanski. *Development of a Coastal Erosion Monitoring Program for the South Shore of Long Island, New York*. Proceedings of a workshop held November 13–14, 1990. New York: Sea Grant Institute.

Dolan, Robert. "Experiences with Atlantic Coast Storms, Barrier Islands, Oregon Inlet, and Douglas Inman." *Shore and Beach* 64, no. 3 (July 1996): 3–7.

Fucella, Joseph E. and Robert Dolan. "Magnitude of Subaerial Beach Disturbance During Northeast Storms." *Journal of Coastal Research* 12, no. 2 (Spring 1996): 420–429.

Griggs, Gary, et al. "The Effects of Storm Waves of 1995 on Beaches Adjacent to a Long-Term Seawall Monitoring Site in Northern Monterey Bay, Calif." *Shore and Beach* 64, no. 1 (January 1996): 34–39.

Hemsley, J. Michael, David W. McGehee, and William M. Kucharski. "Nearshore Oceanographic Measurements: Hints on How to Make Them." *Journal of Coastal Research* 7, no. 2 (Spring 1991): 301–315.

Holman, R. A., and R. T. Guza. "Measuring Run-up on a Natural Beach." *Coastal Engineering* 8 (1984): 129–140.

Holman, R. A., and T. C. Lippmann. "Remote Sensing of Nearshore Bar Systems — Making Morphology Visible." In Nicholas C. Kraus, ed., *Coastal Sediments '87*, New York: American Society of Civil Engineers, 1987.

Howd, Peter A., and William A. Birkemeier. "Storm-Induced Morphology Changes During Duck85." In Nicholas C. Kraus, ed., *Coastal Sediments '87*, New York: American Society of Civil Engineers, 1987.

Kraus, Nicholas C. "The Effects of Seawalls on the Beach: An Extended Literature Review." *Journal of Coastal Research* special issue 4 (Autumn 1988): 1–28.

Kraus, Nicholas C. and William G. McDougal. "The Effects of Seawalls on the Beach: Part 1, An Updated Literature Review." *Journal of Coastal Research* 12, no. 3: 691–701.

Moody, Paul Markert, and Ole Secher Madsen. "Laboratory Study of the Effect of Sea Walls on Beach Erosion." *Massachusetts Institute of Technology Sea Grant Technical Report* 95, no. 3.

Quinn, Mary-Louise. "The History of the Beach Erosion Board, U.S. Army Corps of Engineers, 1930–63."

Sallenger, Asbury H. Jr., et al. "A System for Measuring Bottom Profile, Waves, and Currents in the High Energy Nearshore Environment." *Marine Geology* 51 (1983): 63–76.

U.S. Geological Survey. *National Coastal Geology Program*, 1990.

Weggel, J. Richard. "Seawalls: The Need for Research, Dimensional Considerations, and a Suggested Classification." *Journal of Coastal Research* special issue 4 (Autumn 1988): 29–39.

Wiegel, Robert L. "Coastal Engineering Trends and Research Needs." In Billy L. Edge, ed., *Proceedings of Twenty-First Coastal Engineering Conference*. New York: American Society of Civil Engineers, 1988.

Wise, Randall A., and S. Jarrell Smith. "Numerical Modeling of Storm-Induced Beach Erosion." *The CERCular* 96, no. 1 (March 1996).

Zinn, Donald J. *The Handbook for Beach Strollers*. 1973. Reprint, Chester, Conn.: The Globe Pequot Press, 1973.

9. CONSTITUENCY OF IGNORANCE

Allegood, Jerry. "A Soul-Searching Issue." *Raleigh News and Observer*, April 8, 1996, p. 3A.

——. "Still Fishing for '96 Litter." *Raleigh News and Observer*, February 23, 1998, p. 1A.

Beatley, Timothy. *Ethical Land Use*. Baltimore and London: Johns Hopkins University Press, 1994.

Beatley, Timothy, et al. *Coastal Zone Management*. Washington, D.C.: Island Press, 1994.

Benchley, Peter. *Ocean Planet*. New York: Harry N. Abrams, 1995.

Beston, Henry. *The Outermost House*. New York: Ballantine, 1971.

Coastal Engineering Research Board. *Proceedings of the Sixty-First Meeting of the Coastal Engineering Research Board*. Washington, D.C.: U.S. Army Corps of Engineers, 10 May 1995.

Costanza, Robert, et al. "Valuation and Management of Wetland Ecosystems." *Ecological Economics* 1 (1989): 335–361.

Costanza, Robert, and Charles Perrings. "A Flexible Assurance Bonding System for Improved Environmental Management." *Ecological Economics* 2 (1990): 57–75.

Crowell, Mark, Stephen P. Leatherman, and Michael Buckley. "Historical Shoreline Change: Error Analysis and Mapping Accuracy." *Journal of Coastal Research* 7, no. 3 (Summer 1991): 839–852.

——. "Shoreline Change Rate Analysis: Long Term Versus Short Term Data." *Shore and Beach* (April, 1993): 13–20.

Culliton, Thomas J., et al. *Selected Characteristics of Coastal States, 1980–2000*. National Oceanographic and Atmospheric Administration, October 1989.

Culliton, Thomas J., et al. *Fifty Years of Population Change Along the Nation's Coasts, 1960–2010*. Rockville, Md.: National Oceanic and Atmospheric Administration, April 1990.

Culliton, Thomas J., et al. *Building Along America's Coasts*. Rockville, Md.: National Oceanic and Atmospheric Administration, August 1992.

Denison, Paul S. "Beach Nourishment/Groin Field Constructon Project: Bald Head Island, North Carolina." *Shore and Beach* 66, no. 1 (January 1998): 2–9.

Dolan, Robert. "The Ash Wednesday Storm of 1962: 25 Years Later." *Journal of Coastal Research* 3, no. 2 (1987): ii–vi.

Emerson, Bill, and Ted Stevens. "Natural Disasters: A Budget Time Bomb." *Washington Post*, October 31, 1995, p. A13.

Goudie, Andrew. *The Human Impact on the Natural Environment*. 1981. Reprint, Cambridge: Harvard University Press, 1994.

Griffin, Anna, and Jack Horan. "Teary Residents Shocked at Decimation at Topsail Beach." *Charlotte Observer*, September 8, 1996, p. A16.

Harker's Island, N.C., United Methodist Women. 1987. *Island Born and Bred*. Morehead City, N.C.: Herald Printing, 1991.

Harper, Jim. "BHI to Seek Groins, but Year's Delay Expected." *The [Southport, S.C.] State Port Pilot*, August 31, 1994.

Holing, Dwight, et al. *State of the Coasts*. Washington, D.C.: Coast Alliance, 1995.

Houston, James R. "International Tourism and U.S. Beaches." *The CERCular* 96, no. 2 (June 1996).

Humbach, John A. "Private Property vs. Civic Responsibility." Paper presented at the National Hurricane Conference, Atlantic City, N.J., April 13, 1995.

Institute of Marine and Coastal Sciences, Rutgers the State University of New Jersey. *Coastal Hazard Management Plan for New Jersey*. Summer 1996.

Jones-Jackson, Patricia. *When Roots Die: Endangered Traditions on the Sea Islands*. Athens and London: University of Georgia Press, 1987.

Keene, John C. "Recent Developments in the Law of 'The Takings Issue.' " Paper presented at the National Hurricane Conference, Atlantic City, N.J., April 13, 1995, as revised by the author.

Leatherman, Stephen P., and Jakob J. Møller, "Morris Island Lighthouse: A 'Survivor' of Hurricane Hugo." *Shore and Beach* 59, no. 1 (January 1991): 11–15.

Lisle, Lorance and Robert Dolan. "Coastal Erosion and the Cape Hatteras Lighthouse." *Environ Geol Water Sci* 6, no. 3 (1984): 141–146.

Livermore, S. T. *History of Block Island*. 1877. Reproduced and enhanced by the Block Island Committee of Republication for the Block Island Tercentenary Anniversary, 1961.

Mathews, Jessica. "If a Hurricane Hits Miami." *Washington Post*, October 22, 1996, p. A19.

McHarg, Ian L. *Design With Nature*. Garden City, N.Y.: Doubleday/Natural History Press, 1971

McPhee, John. *Conversations With the Archdruid*. New York: Farrar, Straus and Giroux, 1971.

Monk, John. "Storm Cycle May Make Evacuations Frequent." *Charlotte Observer*, September 8, 1996, p. A1.

National Oceanographic and Atmospheric Administration (Coastal Ocean Office). "Coastal Ocean Program: Science for Solutions." Washington, D.C., 1992.

Navarro, Mireya. "Florida Facing Crisis in Insurance." *New York Times*, April 25, 1996, p. A16.

Owens, David. "The Coastal Management Plan in North Carolina." Raleigh, N.C.: Department of Environmental, Health, and Natural Resources.

Platt, Rutherford. *Land Use and Society*. Washington, D.C.: Island Press, 1996.

Riggs, Stanley R. "Conflict on the Not-So-Fragile Barrier Islands." *Geotimes* 41, no. 12 (December 1996): 14–18.

Scism, Leslie, and Martha Brannigan. "Florida Homeowners Find Insurance Pricey, If They Find It at All." *Wall Street Journal*, July 12, 1996, p A1.

Shipman, Hugh. *Potential Application of the Coastal Barrier Resources Act to Washington State*. Proceedings, Coastal Zone '93, July 19–23, New Orleans, La. New York: American Society of Civil Engineers, 1993, pp. 2243–2251.

Stick, David. *The Outer Banks of North Carolina 1584–1958*. Chapel Hill: University of North Carolina Press, 1958.

Terchunian, Aram V. "Permitting Coastal Armoring Structures: Can Seawalls and Beaches Coexist?" *Journal of Coastal Research* special issue 4 (Autumn 1988): 65–75.

Thompson, Estes (Associated Press). "Hotel at Precarious Site Is Denied a Seawall." *Raleigh News and Observer*, June 27, 1996.

Treaster, Joseph B. "Insurer Curbing Sales of Policies in Storm Areas." *New York Times*, October 1, 1996, p. A1.

Whitlock, Craig. "Fighting Their Watery Fate." *Raleigh News and Observer*, March 6, 1998, p. A1.

Williams, Mark L., and Timothy W. Kana. "Inlet Shoal Attachment and Erosion on Isle of Palms, South Carolina: A Replay." In Nicholas C. Kraus, ed., *Coastal Sediments '87*, pp. 1174–1187. New York: American Society of Civil Engineers, 1987.

Wolverton, Ruthe, and Walt Wolverton. *The National Seashores*. Kensington, Md.: Woodbine House, 1988.

Yeoman, Barry. "Shoreline for Sale." *Coastal Review* 14, no. 2 (Summer 1996): 8–9.

10. FOR SALE

Burn, Billie. *An Island Named Daufuskie*. Spartanburg, S.C.: The Reprint Co., 1991.

Caro, Robert. *The Power Broker: Robert Moses and the Fall of New York*. 1975. Reprint, New York: Vintage, 1975.

De Freese, Duane. "Protecting Coastal Diversity: A Local Perspective." *Coastal Currents* 2, no. 3 (Summer 1994): 12–13.

Di Silvestro, Roger. "Only Fools Build on Shifting Sands." *Audubon* (1989): 106–113.

——. "How a Grandiose Scheme Became a Grand Preserve." *Audubon* (1989): 110.

Dolan, Robert, Bruce P. Hayden, and Gary Soucie. "Environmental Dynamics and Resource Management in the U.S. National Parks." *Environmental Management* 2, no. 3 (1978): 249–258.

Duany, Andres, and Elizabeth Plater-Zyberk. *Towns and Town-Making Principles*. New York: Rizzoli, 1991.

"Dunkirk at South Beach." *Vineyard Gazette*, July 14, 1987.

Ericson, Jody. "Island of Hope." *Nature Conservancy* (January/February 1992): 14–21.

Gay, Jason. "Private Beach Access Costs Are Climbing." *Vineyard Gazette*, July 5, 1996, p. 1D.

Greer, Margaret. *The Sands of Time: A History of Hilton Head Island*. Hilton Head, S.C.: South/Art, 1989.

Griffith, Dorsey. "South Beach Is Sold in Major Chunks to Developers at $2.65 Million Price." *Vineyard Gazette*, July 10, 1987.

——. "Furor Rises on Sale of South Beach. Developers Warned of Legal Action." *Vineyard Gazette*, July 24, 1987, p. 1.

——. "Edgartown Regains South Beach Control in Court Suit to Block Private Developers." *Vineyard Gazette*, July 31, 1987. p. 1.

——. "Town Rejects Offer From Beach Buyers." *Vineyard Gazette*, October 2, 1987, p. 1.

——. "The Story of South Beach: Victory for Town and State." *Vineyard Gazette*, May 12, 1988, p. 1.

Hall, Carla. "Fund-Raising Frenzy Buys Santa Barbara Park." *Los Angeles Times* (Washington edition), March 12, 1996, p. B2.

Hansen, Gunnar. *Islands at the Edge of Time*. Washington, D.C.: Island Press, 1993.

Hayden, B. P., et al. "Long-term Research at the Virginia Coast Reserve." *BioScience* 41, no. 5 (May 1991).

"Houlahan Family Gives Its Position on Sale of Beach." *Vineyard Gazette*, July 17, 1987.

Johnson, Madeleine C. *Fire Island, 1650s–1980s*. Mountainside, N.J.: Shoreline, 1983.

Jones-Jackson, Patricia. *When Roots Die: Endangered Traditions on the Sea Islands*. Athens and London: University of Georgia Press, 1987.

Khoury, Angel Ellis. "Saving Grace: The Resurrection of Corolla's Past." *Outer Banks Magazine* (1996/7): 9–12.

"Land Bank Revenues." *Vineyard Gazette*, July 5, 1996, p. 9E.

"Land Buyers Named." *Vineyard Gazette*, July 31, 1987, p. 11.

"Local Nonprofits as the New Guardians." *California Coast & Ocean* (Summer 1995): 16–17.

Lord, Peter. "Rare Earth: Block Island Wins Recognition as One of Hemisphere's 'Last Great Places.' " *Providence Sunday Journal*, May 12, 1991, p. A1.

McPhee, John. *Encounters with the Arch-Druid*. New York: Noonday: 1971.

Mohney, David, and Keller Easterling, eds. *Seaside: Making a Town in America*. New York: Princeton Architectural Press, 1991.

Ray, G. Carleton, and William P. Gregg Jr. "Establishing Biosphere Reserves for Coastal Barrier Ecosystems." *BioScience* 41, no. 5 (May 1991).

Riddle, Lyn. "As Hilton Head Grows, What of the Environment?" *New York Times*, July 28, 1991, p. 27.

Sinclair, Robert S. "Preserving Paradise." *The Washingtonian* (February 1996): 37–43.

Soper, Tony. *A Natural History Guide to the Coast*. 1984. Reprint, London: Bloomsbury, 1993.

Stick, David. *The Outer Banks of North Carolina*. Chapel Hill: University of North Carolina Press, 1958.

Stolzenburg, William. "Hovering on the Edge." *Nature Conservancy* (March/April 1995): 25–29.

The Trustees of Reservations. "A Guide to the Properties of the Trustees of Reservations." Beverly, Mass. 1992.

Thompson, Marilyn W. "Wild Horses Champ at Tourist Bit." *Washington Post*, August 30, 1996, p. A1.

Thorndike, Joseph J. *The Coast: A Journey Down the Atlantic Shore*. New York: St. Martin's, 1993.

"Tidal Pool of Tears." *Vineyard Gazette*, July 24, 1987.

Twining, Mary A., and Keith E. Baird. *Sea Island Roots*. Trenton, N.J.: Africa World Press, 1991.

Wolverton, Ruthe, and Walt Wolverton. *The National Seashores*. Kensington, Md.: Woodbine House, 1988.